An Introduction to Data Science With Python

As a global academic publisher, Sage is driven by the belief that research and education are critical in shaping society. Our mission is building bridges to knowledge—supporting the development of ideas into scholarship that is certified, taught, and applied in the real world.

Sage's founder, Sara Miller McCune, transferred control of the company to an independent trust, which guarantees our independence indefinitely. This enables us to support an equitable academic future over the long term by building lasting relationships, championing diverse perspectives, and co-creating social and behavioral science resources that transform teaching and learning.

An Introduction to Data Science With Python

Jeffrey S. Saltz

Syracuse University

Jeffrey M. Stanton

Syracuse University

FOR INFORMATION:

2455 Teller Road
Thousand Oaks, California 91320
Email: order@sagepub.com

1 Oliver's Yard
55 City Road
London EC1Y 1SP
United Kingdom

Unit No 323-333, Third Floor, F-Block
International Trade Tower Nehru Place
New Delhi – 110 019
India

18 Cross Street #10-10/11/12
China Square Central
Singapore 048423

Copyright © 2025 by Sage.

All rights reserved. Except as permitted by U.S. copyright law, no part of this work may be reproduced or distributed in any form or by any means, or stored in a database or retrieval system, without permission in writing from the publisher.

All third-party trademarks referenced or depicted herein are included solely for the purpose of illustration and are the property of their respective owners. Reference to these trademarks in no way indicates any relationship with, or endorsement by, the trademark owner.

Printed in Germany
Libri Plureos GmbH
Friedensallee 273
22763 Hamburg

A Library of Congress Cataloging-in-Publishing data record is available for this title

ISBN: 978-1-0718-5065-7

This book is printed on acid-free paper.

Acquisitions Editor: Adeline Grout

Editorial Assistant: Cassie Carey

Production Editor: Laura Barrett

Copy Editor: Karin Rathert

Typesetter: TNQ Tech Pvt. Ltd.

Indexer: TNQ Tech Pvt. Ltd.

Cover Designer: Scott Van Atta

Marketing Manager: Victoria Velasquez

24 25 26 27 28 10 9 8 7 6 5 4 3 2 1

BRIEF CONTENTS

Instructor Preface		xi
Acknowledgments		xiii
About the Authors		xv
Introduction: Data Science, Many Skills		1
Chapter 1	Begin at the Beginning With Python	9
Chapter 2	Rows and Columns	23
Chapter 3	Data Munging	41
Chapter 4	What's My Function?	59
Chapter 5	Beer, Farms, Peas, and Statistics	75
Chapter 6	Sample in a Jar	93
Chapter 7	Storage Wars	109
Chapter 8	Pictures Versus Numbers	129
Chapter 9	Map Magic	157
Chapter 10	Linear Models	179
Chapter 11	Classic Classifiers	199
Chapter 12	Left Unsupervised	223
Chapter 13	Words of Wisdom: Doing Text Analysis	243
Chapter 14	In the Shallows of Deep Learning	263
Bibliography		**283**
Index		**285**

DETAILED CONTENTS

Instructor Preface — xi
Acknowledgments — xiii
About the Authors — xv

Introduction: Data Science, Many Skills — 1
- What is Data Science? — 2
- The Steps in Doing Data Science — 3
- The Skills Needed to Do Data Science — 4
- Identifying Data Problems Through Stories — 5
- Case: Overall Context and Desired Actionable Insight — 7
- Chapter Challenges — 8

Chapter 1   Begin at the Beginning With Python — 9
- Getting Ready to Use Python — 10
- Using Python in a Jupyter Notebook — 11
- Creating and Using Lists — 12
- Slicing Lists — 15
- The Virtual Machine — 16
- Shared Python Code Libraries: The Package Index — 17
- Chapter Challenges — 22

Chapter 2   Rows and Columns — 23
- Creating Pandas DataFrames — 25
- Exploring DataFrames — 28
- Accessing Columns in a DataFrame — 30
- Accessing Specific Rows and Columns in a DataFrame — 33
- Generating DataFrame Subsets With Conditional Evaluations — 34
- A Quick Review — 36
- Chapter Challenges — 39

Chapter 3   Data Munging — 41
- Reading Data From a CSV Text File — 42
- Removing Rows and Columns — 44

Renaming Rows and Columns	46
Cleaning Up the Elements	47
Sorting and Grouping DataFrames	49
Grouping Within DataFrames	51
Chapter Challenges	56

Chapter 4   What's My Function? — 59

Why Create and Use Functions?	60
Creating Functions in Python	61
Defensive Coding	64
Classes and Methods	67
Chapter Challenges	73

Chapter 5   Beer, Farms, Peas, and Statistics — 75

Historical Perspective	76
Sampling a Population	77
Understanding Descriptive Statistics	77
Using Descriptive Statistics	80
Using Histograms to Understand a Distribution	83
Normal Distributions	87
Chapter Challenge	90

Chapter 6   Sample in a Jar — 93

Sampling in Python	95
A Repetitious Sampling Adventure	96
Law of Large Numbers and the Central Limit Theorem	99
Making Decisions With a Sampling Distribution	100
Evaluating a New Sample With Thresholds	102
Chapter Challenges	107

Chapter 7   Storage Wars — 109

Accessing Excel Data	111
Working With Data From External Databases	112
Accessing a Database	113
Accessing JSON Data	117
Chapter Challenges	126

Chapter 8   Pictures Versus Numbers — 129

A Visualization Overview	130

Basic Plots in Python	132
Using Seaborn	134
Scatterplot Visualizations	141
Chapter Challenges	154

Chapter 9   Map Magic — 157

Map Visualizations Basics	158
Creating Map Visualizations With Folium	160
Showing Points on a Map	166
Chapter Challenges	177

Chapter 10   Linear Models — 179

What is a Model?	180
Supervised and Unsupervised Learning	180
Linear Modeling	181
An Example—Car Maintenance	183
Partitioning Into Training and Cross Validation Datasets	191
Using K-Fold Cross Validation	193
Chapter Challenges	197

Chapter 11   Classic Classifiers — 199

More Supervised Learning	199
A Classification Example	200
Supervised Learning With Naïve Bayes	205
Naïve Bayes in Python	206
Supervised Learning Using Classification and Regression Trees	211
Chapter Challenges	220

Chapter 12   Left Unsupervised — 223

Supervised Versus Unsupervised	224
Data Mining Processes	224
Association Rules Data	225
Association Rules Mining	226
How the Association Rules Algorithm Works	233
Visualizing and Screening Association Rules	235
Chapter Challenges	242

Chapter 13 Words of Wisdom: Doing Text Analysis   243

    Unstructured Data   244
    Reading in Text Files   245
    Creating the Word Cloud   246
    Sentiment Analysis   248
    Topic Modeling   251
    Other Uses of Text Mining   257
    Chapter Challenges   262

Chapter 14 In the Shallows of Deep Learning   263

    The Impact of Deep Learning   264
    How Does Deep Learning Work?   264
    Deep Learning in Python—a Basic Example   267
    Deep Learning Using the MNIST Data   272
    Chapter Challenges   280

Bibliography   283

Index   285

INSTRUCTOR PREFACE

Data science is now well established as an area of study at many universities, with courses and academic programs in data science offered at institutions around the world. In addition, the skills of data science hold increasing value for students from many disciplines, including social sciences and the humanities. Given the growing importance and popularity of data science, our goal in creating this book was to create an easy-to-understand, introductory-level text that offers intuitive approaches to applying data science methods across a variety of contexts.

Given the variety of backgrounds that students may bring to the classroom, we have tried to make this book as engaging, straightforward, and readable as possible. Students whose mathematical preparation has included high school algebra will find the concepts and code quite intelligible. The hands-on examples provided—including the case study that serves as a common thread through all of the chapters—will also be relatable to most students. We are confident of this because we use these examples, code, and data sets in our own introductory data science course—a course that has been refined and improved at Syracuse University for more than a decade. Regardless of the department from which your introductory data science course is offered, the use of this book will give your course broad appeal to students from across your institution.

For this edition of the book, we have chosen Python and the Jupyter Notebook environment as the teaching and learning platforms. Chapter 1 provides a compact introduction to the language without intending to duplicate the many excellent novice resources available for Python learners. (Instructors who prefer teaching in R may want to consider *Data Science for Business with R*.) Python is a free and open-source language designed to be easy for nonprogrammers to learn and offering a robust ecosystem of packages for analysis, visualization, and data management. Many professional data scientists consider Python as the primary language of data science. Instructors can use any Python development environment that they prefer, but the use of Jupyter Notebooks makes it possible for instructors and students to create and modify Python code using virtually any web browser without having to install any software. As a benefit to students and instructors, we have provided one standalone Jupyter Notebook file for each chapter—you can find links to these notebooks on the publisher's companion website along with sample slide decks and other instructional material. All of the datasets used in the notebooks are available from public repositories.

New in this edition is a concluding chapter on deep learning, the basis of many recent and startling advances in machine learning and artificial intelligence. Using nontechnical language, the new chapter introduces the essential ideas behind deep learning and develops a simple predictive model to illustrate the training process. Also new is a unified bibliography following the concluding chapter—this provides many suggestions for follow-up readings.

TEACHING RESOURCES

This text includes an array of instructor teaching materials designed to save you time and help you keep students engaged. To learn more, visit **sagepub.com** or contact your Sage representative at **sagepub.com/findmyrep.**

ACKNOWLEDGMENTS

Many thanks to Leah Fargotstein, Jennifer Milewski, and the great team of folks at Sage Publications, who ably assisted us with the development of this book and its previous editions. We would also like to acknowledge our colleagues Syracuse University's School of Information Studies, who helped us gather student feedback to improve this book and Akit Kumar, who double-checked the accuracy of our code and instructions.

There were a number of reviewers we would like to thank who provided extremely valuable feedback during the development of the manuscript:

- John Bono, *University of Maryland, College Park*
- Yi Liu, *University of the Incarnate Word*
- James N. Maples, *Eastern Kentucky University*
- Minjuan Wang, *San Diego State University*

ABOUT THE AUTHORS

Jeffrey S. Saltz is an associate professor at Syracuse University in the School of Information Studies and director of the school's master's of science program in applied data science. His research and teaching focus on helping organizations leverage information technology and data for competitive advantage. Specifically, his current research focuses on the socio-technical aspects of data science projects, such as how to coordinate and manage data science teams. To stay connected to the "real world," Saltz consults with clients ranging from professional football teams to Fortune 500 organizations. Prior to becoming a professor, Saltz's two decades of industry experience focused on leveraging emerging technologies and data analytics to deliver innovative business solutions. In his last corporate role, at JPMorgan Chase, he reported to the firm's chief information officer and drove technology innovation across the organization. Saltz also held several other key technology management positions at the company, including chief technology officer and chief information architect. He also served as chief technology officer and principal investor at Goldman Sachs, where he helped incubate technology start-ups. He started his career as a programmer, project leader, and consulting engineer with Digital Equipment Corp. Saltz holds a BS degree in computer science from Cornell University, an MBA. from The Wharton School at the University of Pennsylvania, and a PhD in Information Systems from the New Jersey Institute of Technology.

Jeffrey M. Stanton, PhD, is a professor at Syracuse University in the School of Information Studies. Stanton's research focuses on the impacts of machine learning on organizations and individuals. He is the author of *Reasoning With Data* (2017), an introductory statistics textbook. Stanton has also published many scholarly articles in peer-reviewed behavioral science journals, such as the *Journal of Applied Psychology, Personnel Psychology*, and *Human Performance*. His articles also appear in *Journal of Computational Science Education, Computers and Security, Communications of the ACM, Computers in Human Behavior*, the *International Journal of Human-Computer Interaction, Information Technology and People*, the *Journal of Information Systems Education*, the *Journal of Digital Information, Surveillance and Society, and Behaviour & Information Technology*. He also has published numerous book chapters on data science, privacy, research methods, and program evaluation. Stanton's research has been supported through 19 grants and supplements including the National Science Foundation's CAREER award. Before getting his PhD, Stanton was a software developer who worked at startup companies in the publishing and professional audio industries. He holds a bachelor's degree in computer science from Dartmouth College and a master's and PhD in psychology from the University of Connecticut.

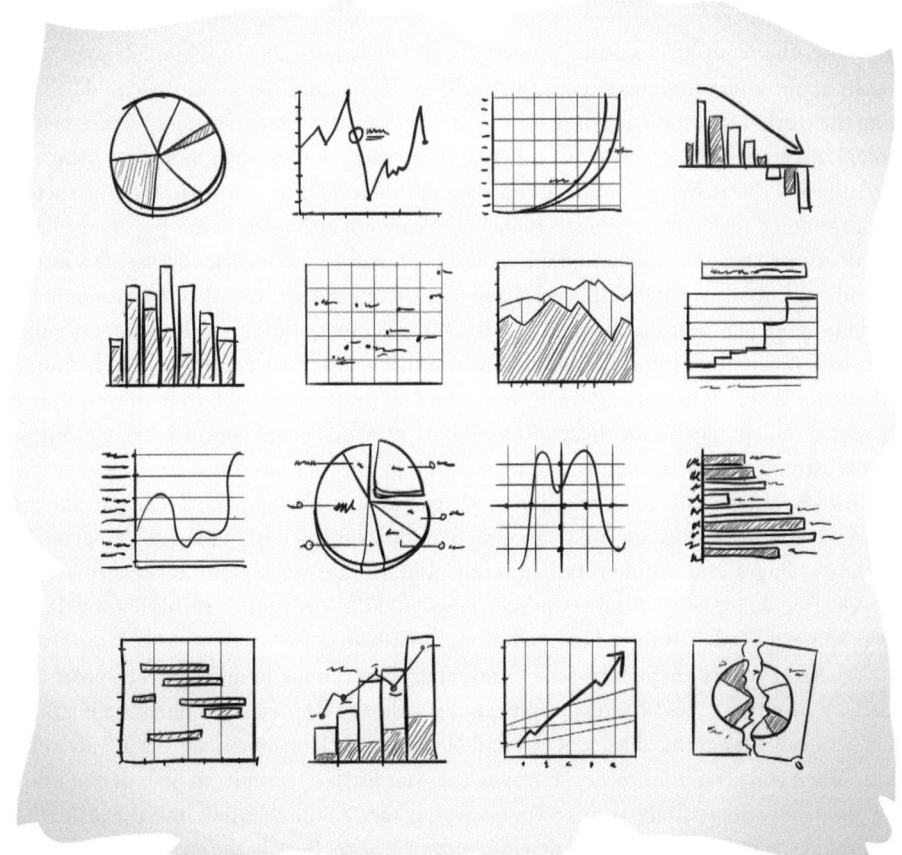

INTRODUCTION: DATA SCIENCE, MANY SKILLS

LEARNING OBJECTIVES

Articulate what data science is.

Describe the applicability of data science to business.

Understand the steps, at a high level, of doing data science.

Describe the roles and skills of a data scientist.

WHAT IS DATA SCIENCE?

For some, the term "data science" evokes images of statisticians in white lab coats staring fixedly at blinking computer screens filled with scrolling numbers. Nothing could be farther from the truth. First, statisticians do not wear lab coats: this fashion statement is reserved for biologists, physicians, and others who have to keep their clothes clean in environments filled with unusual fluids. Second, much of the data in the world is nonnumeric and unstructured. In this context, unstructured means that the data are not arranged in neat rows and columns. Think of a web page full of photographs and short messages among friends: very few numbers to work with there. Although it is certainly true that businesses use plenty of numeric information—sales of products, financial reports, and customer satisfaction are a few examples—there is so much other information in the world that mathematicians look at and cringe. So, although it is always useful to develop your math skills, there is much to be accomplished in the world of data science for those of us who are presently more comfortable working with words, lists, photographs, sounds, and other kinds of information.

In addition, data science is much more than simply analyzing data. There are many people who enjoy analyzing data and who could happily spend all day looking at histograms, calculating averages, and running statistical tests, but for those who prefer other activities, data science offers a range of roles and requires a range of skills. Let's consider this idea by thinking about some of the data involved in buying a box of cereal.

Whatever your cereal preferences—fruity, chocolaty, fibrous, or nutty—you prepare for the purchase by writing "cereal" on your grocery list. Already your planned purchase is a piece of data, also called a datum, albeit a pencil scribble on the back on an envelope that only you can read. When you get to the grocery store, you use your list as a reminder to grab that jumbo box of FruityChocoBoms off the shelf and put it in your cart. At the checkout line, the cashier scans the barcode on your box, and the cash register logs the price. Back in the warehouse, a computer tells the stock manager that it is time to request another order from the distributor because your purchase was one of the last boxes in the store. You also have a coupon for your big box, and the cashier scans that, giving you a manufacturer discount. At the end of the week, a report of all the scanned manufacturer coupons gets uploaded to the cereal distributor so they can issue a reimbursement to the grocery store for all of the coupon discounts they have given to customers. Finally, at the end of the month a store manager looks at a colorful collection of pie charts showing the different kinds of cereal that were sold and, on the basis of strong sales of fruity cereals, decides to offer more varieties of these on the store's limited shelf space next month.

The small piece of information that began as a scribble on your grocery list ended up in many different places but most notably on the desk of a manager as an aid to decision-making. On the trip from your pencil to the manager's desk, your list item went through many transformations. In addition to the computers where the item might have stopped by or stayed on for the long term, many other pieces of hardware—such as the barcode scanner—were involved in collecting, manipulating, transmitting, and storing your purchase data. In addition, many different pieces of software were used to organize, aggregate, visualize, and present the data collected from you and other customers. Finally, many different human systems were involved in working with the data. People decide which systems to buy and install, who should get access to what kinds of data, and what would happen to the data after its immediate purpose was fulfilled. The personnel of the grocery chain and its partners made a thousand other detailed decisions and negotiations before the scenario described here could become reality.

THE STEPS IN DOING DATA SCIENCE

Obviously, data scientists are not involved in all of these steps. Data scientists don't design or build barcode readers, for instance. So where would the data scientists play the most valuable role? Generally speaking, data scientists play the most active roles in the four As of data: data architecture, data acquisition, data analysis, and data archiving. Using our cereal example, let's look at these roles one by one. First, with respect to architecture, it was important in the design of the point-of-sale system (what retailers call their cash registers and related gear) to think through in advance how different people would make use of the data coming through the system. The system architect, for example, had a keen appreciation that both the stock manager and the store manager would need to use the data scanned at the registers, albeit for somewhat different purposes. A data scientist would help the system architect by providing input on how the data would need to be routed and organized to support the analysis, visualization, and presentation of the data to the appropriate people.

Next, acquisition focuses on how the data are collected and, importantly, how the data are represented prior to analysis and presentation. For example, each barcode represents a number that, by itself, is not descriptive of the product it represents. At what point after the barcode scan, does the product identification number become associated with a description of the product or its price or its net weight or its packaging type? Different barcodes are used for the same product (e.g., for different-sized boxes of cereal). When should we make note that purchase X and purchase Y are the same product but in different packages? Representing, transforming, grouping, and linking the data are all tasks that need to occur before the data can be profitably analyzed, and these are all tasks in which the data scientist is actively involved.

The analysis phase is another area data scientists are heavily involved in. In this context, we are including summarization of the data, using subsets of data (samples) to make inferences about the larger context, and visualizing of the data by presenting it in tables, graphs, and even animations. Although there are many technical, mathematical, and statistical aspects to these activities, keep in mind that the ultimate audience for the results of a data analysis is always a person or people. These people are the data users, and fulfilling their needs is the primary job of a data scientist. This point highlights the need for excellent communication skills in data science. The most sophisticated statistical analysis ever developed will be useless unless the results can be effectively communicated to the data user.

Finally, the data scientist must become involved in the archiving of the data. Preservation of collected data in a form that makes it highly reusable—what you might think of as data curation—is a challenge because it is so hard to anticipate all of the future uses of the data. For example, when the developers of X were working on how to store tweets, they probably never anticipated that tweets would be used to pinpoint earthquakes and tsunamis, but they had enough foresight to realize that geocodes—data that show the geographical location from which a tweet was sent—could be a useful element to store with the data.

THE SKILLS NEEDED TO DO DATA SCIENCE

All in all, our cereal box and grocery store example helps highlight where data scientists get involved and the skills they need. Here are some of the skills that the example suggested:

Learning the application domain: The data scientist must quickly learn how data will be used in a particular context.

Communicating with data users: A data scientist must possess skills for learning the needs and preferences of data users. The ability to translate back and forth between computing or statistics and the vocabulary of business is a critical skill.

Seeing the big picture of a complex system: After developing an understanding of the application domain within a business, the data scientist must imagine how data will move around among all of the relevant systems and people.

Knowing how data can be represented: Data scientists must have a clear understanding about how data can be stored and linked as well as about metadata (data that describe how other data are arranged).

Data transformation and analysis: When data become available for the use of managers, data scientists must know how to transform, summarize, and make inferences from the data. As noted, being able to communicate the results of analyses to data users is also a critical skill here.

Visualization and presentation: Although numbers often have the edge in precision and detail, a good data display (e.g., a bar chart) can often be a more effective means of communicating results to data users.

Attention to quality: No matter how good a set of data might be, there is no such thing as perfect data. Data scientists must know the limitations of the data they work with, know how to quantify its accuracy, and be able to make suggestions for improving the quality of the data in the future.

Ethical reasoning: If data are important enough to collect, they are often important enough to affect the lives of employees, customers, and others. Data scientists must understand important ethical issues, such as privacy, and must communicate about the nature of data to try to prevent misuse of data or analytical results.

The skills and capabilities noted are the tip of the iceberg, of course, but notice what a wide range is represented here. Although a keen understanding of numbers and mathematics is valuable, particularly for data analysis, the data scientist also needs to have excellent communication skills, be a great systems thinker, have an eye for visual data displays, and be highly capable of thinking critically about how data will be used to make decisions and affect people's lives. Of course, there are few people who are good at all of these things, so some of the people interested in data science will specialize in one area, whereas others will become experts in another area. This highlights the importance of teamwork as well.

IDENTIFYING DATA PROBLEMS THROUGH STORIES

When we eat some cereal from a box we bought at the grocery store, we must ultimately thank farmers for producing all of the foodstuffs that went into that box. Depending upon our cereal preferences, those ingredients might include wheat, oats, raisins, nuts, and/or honey, among other crops. And don't forget the milk that you poured over your cereal!

Farmers are highly dependent on the natural environment when producing the food we consume. A late spring frost can kill seedlings or blossoms. Hail or wind in the summer can damage the stalks and fruit. In this highly physical world of unpredictable natural forces, what role is there for data science? Having a nose for identifying data-related problems in unfamiliar domains requires openness, curiosity, creativity, and a willingness to ask many questions. Most data scientists must (eventually) become immersed in the problem domain where they are working. The data scientist will probably not become a farmer, but if you are going to identify a data problem that a farmer has, you have to learn to think like a farmer to some degree.

To get this domain knowledge, you can read or watch videos, but the best way is to ask subject matter experts (SMEs) about what they do. The whole process of asking questions deserves its own treatment, but for now there are three issues data scientists consider when asking questions. First, you want SMEs to tell stories about what they do. Then you want to ask them about anomalies: the unusual things that happen for better or for worse. Finally, you want to ask about risks and uncertainty: about the situations where it is hard to tell what will happen next, when what happens next could have a profound effect on whether the situation ends badly or well. Each of these three areas of questioning reflects an approach to identifying data problems that might turn up a task or solution that could be accomplished with data, information, and the right decision at the right moment.

The purpose of asking SMEs to tell stories is that people mainly think in stories. From farmers to teachers to managers to CEOs, people know and tell stories about success and failure in their particular domains. Stories are powerful ways of communicating wisdom among different members of the same profession, and they are ways of collecting a sense of identity that sets one profession apart from another. The only caution is that stories can contain inaccuracies. As data scientists, we listen to the stories, and then we collect data to help verify them.

If you can get a professional to tell the main stories that guide how they conduct their work, you must consider how to verify those stories. Without questioning the veracity of the person who tells the story, you can imagine ways of measuring the different aspects of how things happen in the story with an eye toward eventually verifying (or sometimes debunking) the stories that guide professional work. For example, a farmer might say that in the deep spring frost that occurred 5 years ago, the trees in the hollow were spared frost damage, whereas the trees around the ridge were not. For this reason, on a cold night, the farmer places smudge pots (containers that create a smoky fire) around the ridge. The farmer believes that this strategy works, but does it? You could get a few inexpensive digital weather stations and record data in the fields on cold and warm nights. You could use this data to create a model of temperature changes in the different areas, and this model could support, revise, or refute the story.

A second strategy for problem identification is to seek out exception cases. Exceptions may seemingly be created by chance factors, or they may be caused by human intervention: We are interested in both. Later in this book we will learn about how the classic methods of statistical inference characterize the average—the most typical cases that occur—and then examine extreme cases that may lie far from the center. Identifying unusual cases is a powerful way of understanding how things work, but it is necessary first to define the central or most typical occurrences before one can have an accurate idea of what constitutes an unusual case. Statistical inference is an important tool in the data scientist's tool kit for drawing a distinction between the usual and the unusual.

A third strategy for identifying data problems is to uncover the sources and effects of risk and uncertainty. A basic function of accurate information is to reduce uncertainty. It is often valuable to reduce uncertainty because of how risk affects the things we all do. At work, at school, and at home, life is full of risks: making a decision (or failing to make one) sets off a chain of events that could lead to something good or something not so good. We generally want to narrow down the possibilities in a way that maximizes the chances of a good outcome and minimizes the chances of a bad one. To do this, we need to make better decisions, and to make better decisions we need to reduce uncertainty. By asking questions about risks and uncertainty, a data scientist zeroes in on the situations that matter. You can even look at the two previously discussed strategies—asking about the stories that comprise professional wisdom and asking about anomalies or unusual cases—in terms of their potential for reducing uncertainty and risk.

For farmers, considerable risk comes from the weather, and the uncertainty revolves around which countermeasures will be most cost-effective at protecting the crops. Consuming lots of oil in smudge pots on warm nights is a waste of resources that could make the difference between a profitable or an unprofitable year. Precise and timely information about local weather conditions could also become a valuable area for problem-solving with data. What if a livestream of Doppler radar could appear on the farmer's smartphone? The app could provide predicted wind speed and temperature for the farm, and a statistical model provided by a data scientist could compute the likelihood of damage.

Of course, there are many other situations where data science could prove useful. For example, banks have used data science for many years to perform credit analysis when a customer wants to take out a loan or obtain a credit card. Retailers use data science to try to predict inventory and to model losses due to theft. Online marketers use data science to group people into clusters so that an advertisement can suggest a related product to someone who liked a certain product (such as a movie). Finally, smart devices in our homes, such as a digital thermostat, use data to save money on energy costs. Although it would take an entire book to describe the many different situations where data science has been or could be used, hopefully these examples give you a feel for what is possible when data science is applied to real-world challenges.

To recap, there are many different contexts in which a data scientist might work, and doing data science requires much more than sitting in front of a computer and coding. The data scientist needs to understand the domain and data in that domain. The data scientist can obtain this knowledge by talking to or observing SMEs. One strategy for problem identification is to interact with an SME and get that person to tell a story about their domain. A second strategy is to look for good and bad exceptions. Finally, a third strategy is to explore risk and uncertainty.

CASE: OVERALL CONTEXT AND DESIRED ACTIONABLE INSIGHT

In this book, we use a large dataset from an airline to explore and demonstrate some the skills and capabilities needed by data scientists. In each chapter, we open this customer satisfaction data, describe the kinds of challenges that businesses such as airlines want to address, and then use the data as a demonstration of how data science might help. The open-source programming language known as Python and an interactive graphical web-based "notebook" environment are what we use to work with real data examples to illustrate both the challenges of data science and some of the techniques used to address those challenges.

Note that the idea of "big data" is a closely related area of focus. In brief, big data is data science that is focused on processing and analysis of very large datasets. Of course, there's no particular size threshold that defines a "very large dataset," but for our purposes we define big data as trying to analyze datasets that are so large that one could not address the challenge using the computing power of a laptop or other personal computer. As an example of a big data problem to be solved, we know that one large retailer adjusts its pricing in near real time for 73 million item, based on demand and inventory. As you might guess, the amount of data and calculations required for this type of analysis is too large for a laptop. However, the techniques covered in this book are conceptually similar to how one would approach that kind of problem. Several chapters in the book consider different aspects of big data concepts and strategies.

No one book can cover the wide range of activities and capabilities involved in a field as diverse and broad as data science. Throughout the book, references to other guides and resources provide the interested reader with access to additional information. In the open source spirit of Python, these are, wherever possible, web based and free. One of the resources we recommend is Wikipedia, the free, online, user-sourced encyclopedia. As a collectively developed and edited encyclopedia, Wikipedia is admittedly imperfect, but it can be a useful learning resource. You can't become an expert on a topic by consulting only Wikipedia, but you can certainly become smarter by starting there.

Another useful resource is Khan Academy. Most people think of Khan Academy as a set of videos that explain math concepts to middle and high school students, but thousands of adults around the world use Khan Academy as a refresher course for a range of topics or as a quick introduction to a topic that they never studied before. All of the lessons at Khan Academy are free, and if you log in with a Google or Facebook account, you can do exercises and keep track of your progress.

One last thing: The book presents topics in an order that should work well for people with little or no prior experience in areas like computer science or statistics. If you already have knowledge, training, or experience in computer science and/or statistics, you should feel free to skip over some of the introductory material and move right into the topics and chapters that interest you most.

CHAPTER CHALLENGES

1. To help structure discussions with SMEs, an interview guide is useful. Create an interview guide to ask questions of an SME. Try to create one that is general purpose, and then refine it so that you can use it for different types of SMEs.

2. Refine your SME interview guide from the previous item so that it includes questions that a hotel manager might answer.

3. Make a list of five of the main risks that a hotel manager faces. It might help to give some thought to a few of the major issues that hotel guests find problematic.

4. Choose any one of the risks that you identified for the previous item. What kind of data would you collect that could help you understand and possibly reduce that risk?

iStock.com/Nynke van Holten

1 BEGIN AT THE BEGINNING WITH PYTHON

LEARNING OBJECTIVES

Know how to get started with Python.

Gain familiarity with using Jupyter notebooks.

Build and manipulate data objects, such as lists, in Python.

If you are new to computers, programming, and/or data science, welcome to an exciting world that will open the door to the most powerful programming language for data analytics ever created anywhere in the universe—no joke. Contrastingly, if you are experienced with spreadsheets, statistical analysis, or accounting software, you are probably thinking, as you start this chapter, that using a programming language to do data science is excessive. Both perspectives are reasonable. The Python open-source programming language is immensely powerful, flexible, and extensible (meaning that people can create new capabilities for it quite easily). At the same time, Python is a complete and capable programming language, meaning that most of the work that one needs to perform is done through carefully crafted text instructions, many of which have specific syntax (the punctuation and related rules for creating a line of code that works). Additionally, like many programming languages, Python is not always great at giving feedback or error messages that help the user fix mistakes or figure out what is wrong when results look weird.

But there is a method to the madness here. One of the virtues of Python as a teaching tool is that it hides little. A successful user must fully understand what the data situation is or else the Python code will not work. With a spreadsheet, it is easy to type in a lot of numbers and a formula like =FORECAST, and a result pops into a cell like magic, whether the results make any sense or not. With Python you have to know your data, what you can do with it, whether and how it must be transformed, and how to check for trouble. Because Python is a programming language, it forces users to think about problems in terms of data objects, attributes of those data objects, and methods that can be applied to those objects. These are important metaphors used in modern programming languages, and no data scientist can succeed without having at least a rudimentary understanding of how code is programmed, tested, and integrated into working systems.

The extensibility of Python means that new libraries are being added all the time by volunteers: for example, Python has dozens of libraries related to using neural networks for deep learning. You can be sure that, whatever the next big development is in the world of data, someone in the Python community will start to develop a new code library that will make use of it.

Finally, the lessons we learn by working with Python have nearly universal applicability to other programs and environments. If you have mastered Python, it is a reasonably small step to get the hang of R, another important data science language. Because Python is open source, there are no licensing fees paid by schools, students, or teachers. As a result, it is possible to learn the most powerful data analysis language in the universe for free and take those skills and knowledge with you no matter where you go. It will take some patience, though, so please hang in there!

GETTING READY TO USE PYTHON

Let's get started. Obviously, you will need a computer. If your computer has the Windows(r), Mac-OS-X(r), or a Linux operating system, there is a version of Python available for it. There are two major versions of Python in circulation: we will be using the newer one, known as

Python 3, exclusively. One valuable place to learn more about Python is the official site of the Python Software Foundation at https://www.python.org. The website includes a beginner's guide and a list of books about learning Python.

We are going to save ourselves a lot of hassle, however, by not installing Python but instead using it in a browser-based environment known as Jupyter notebooks (the name "Jupyter" refers to the three main programming languages that are supported: Julia, Python, and R). The notebook environment for coding has become immensely popular over recent years, particularly for new language learners. A notebook is conceptually simple—it is a web page divided into sections called "cells," where each cell can be either a piece of operational code or some documentation. The amazing thing about notebooks is that the code you write in the code cells can be run and tested right in your web browser. You do not need to install any new software on your computer. The use of a web page to organize your code also means that you can view, run, modify, and test your code on a mobile device. You can also share your notebook with another person.

For this book, we are going to use Jupyter notebooks exclusively. There are many different available sites for running notebooks, and some colleges and universities even offer their own notebook environments. You can use any Jupyter environment you like, but in this book, we will be illustrating techniques and examples using free notebooks offered by Google in its Colab environment. Colab offers a specially enhanced version of Jupyter notebooks with many tools to help coders. The Colab environment includes free use of a virtual machine hosted by Google to run your code. Although the virtual machine has somewhat limited memory and disk space, for virtually everything we do in this book, those limits are not a problem. To get started with Colab you will need a Google Drive account. If you have a Gmail address, then you already have access to Drive. Otherwise sign up for one now, and then go to https://drive.google.com to get started. When you are in Google Drive, pull down the New menu, and click on "More." You will have the option to "Connect More Apps," and if you pick this, you will be able to search for "Colaboratory" and attach that to your account. After that, you will be able to pull down New and More and create a new Colab notebook. New notebooks default to the Python language, so you will be able to start coding immediately.

As a companion resource to this book, we have provided Colab notebooks to go with every chapter. Look on the publisher website for a link. These notebooks will be updated from time to time to keep up with changes in the Python language and the additional extensions that we use for data science tasks.

USING PYTHON IN A JUPYTER NOTEBOOK

The screenshot shows a simple command to type that shows the most basic method of interaction with Python in the notebook environment. Notice that a new notebook comes with one empty code cell already created. There's a large "play" button to the left of the code cell that lets you run your code when you are ready. In the screen shot, the user has typed "1+1" and is ready to click on play. As soon as we click the play button, Python will dutifully report the

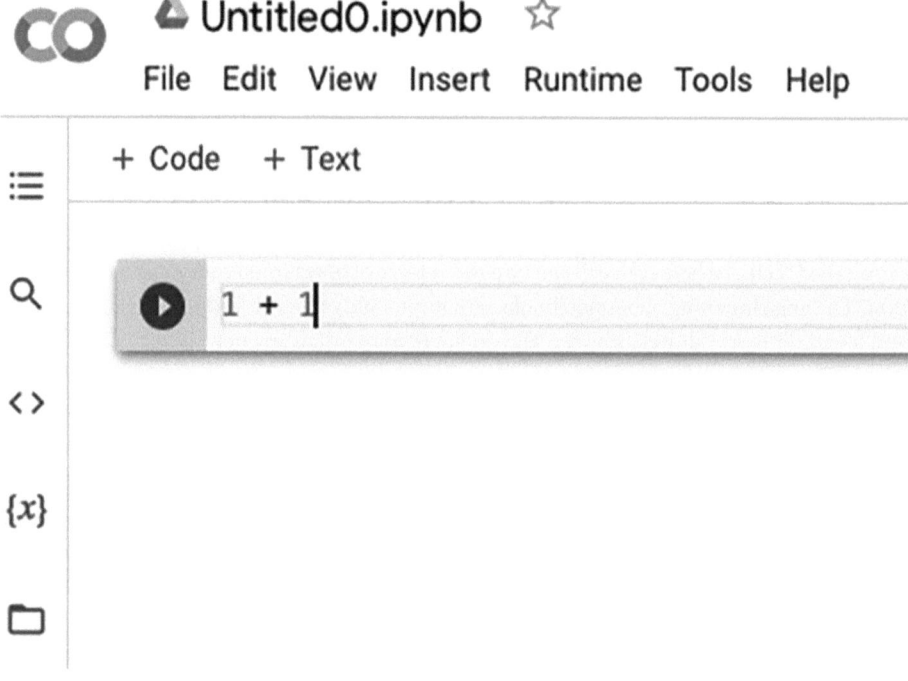

result as 2. Note that the first time you press the play button, there will be few seconds of delay while your virtual machine starts.

You can add more code to the same cell and/or create another code cell. As a general rule, only the last line of code in a particular cell produces visible output. In Colab, you can use the Insert menu to add a code cell, or there is a "+ Code" button that appears if you hover near the center-top or center-bottom of any existing cell.

CREATING AND USING LISTS

In the realm of data science, we often have a need to store and manipulate a list of data points. Python provides highly useful data structures for this including one that is known as a list. We suggest that you follow along with this and other exercises in the book by typing and running the suggested code in a notebook or by following along in the notebooks we have provided. To build a list, first create a new code cell. Here is a list of the ages of members of a family: 43, 42, 12, 8, 5, for Dad, Mom, Sis, Bro, and their Dog, respectively. This is a list of items, all of the same type, specifically an integer. These are integers because there are no decimal points and therefore nothing after the decimal point. We can create a list of integers in Python using square brackets. We can also create a list of strings (character

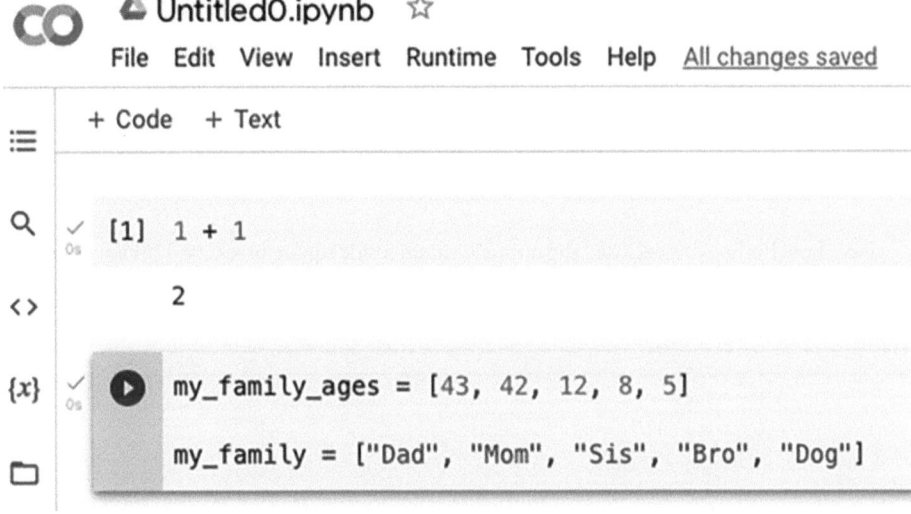

data) using the same method but putting each item in quotes. Take a look at the screen shot: The first line of code in this example creates a list of integers, and the second line creates a list of strings.

This is the last time that a complete screenshot from Python will appear in the book. From here on out we will look at lines of code and pieces of output so that we don't waste so much space on the page. If you are using an e-book, you might also be able to copy and paste lines of code from the e-book into a code cell of a notebook. In these two lines of code, we have assigned all of the stuff that appears to the right of the equals sign into the variable that is named to the left of the equals sign. Here's the same code as plain text:

```
my_family_ages = [43, 42, 12, 8, 5]

my_family = ["Dad", "Mom", "Sis", "Bro", "Dog"]
```

Note that the names of the variables consist of one or two words separated by an underscore. This is a common way of naming a variable in Python. Also note that the quote marks used in the second list must be the simple ones that go straight up and down rather than the slanted smart quotes that some word processing programs provide. Because this is a two-color book, we will show the code in blue and any comments or output in black. You will notice in your notebook environment that the Colab color-codes the different elements of the code you type. For example, the integers may be shown in green and the strings in red. This color scheme helps avoid or solve programming errors that occur as a result of punctuation problems and other issues. If you have the e-book and you copy and paste something from the book into Colab, check the result carefully to make sure that nothing has gotten lost in translation. As noted, you may find it easier to work from copies of the notebook files we have provided.

In a code cell, we can type the name of a variable, and as long as it is the last code statement in a cell, Python will report what that variable contains when we run that cell:

```
my_family_ages

[43, 42, 12, 8, 5]
```

Note how Python has enclosed the numeric values inside square brackets: This is a visual confirmation from Python that the data are stored in a list. If we want a code cell to produce more than one chunk of output, we should use the print() function to display the data that we want to see:

```
print(my_family_ages)

print(my_family)

[43, 42, 12, 8, 5]

['Dad', 'Mom', 'Sis', 'Bro', 'Dog']
```

Each call to print() starts a new line of output, so the contents of our two variables are neatly shown on two lines. "Echoing" a data object to make its contents visible in your Python notebook is a simple but important tool. Any time you want to know what is in a data object in Python, type the name of the object or print() it, and Python will report it back to you. In the next command, we begin to see the power of Python:

```
sum(my_family_ages)

110
```

This code asks Python to add together all of the numbers in my_family_ages, which turns out to be 110 (you can check it yourself with a calculator if you want). This is perhaps an odd thing to do with the ages of family members, but it shows how with a short and simple line of code, you can unleash quite a lot of processing on your data. We could, for example, calculate the mean age of our family members by dividing the sum by the number of observations, which in this case is the same thing as saying the length of the list:

```
sum(my_family_ages)/len(my_family_ages)

22.0
```

Finally, for fun, let's try to call a function called fish() that doesn't exist with this line of code: fish(my_family_ages). Pretty much as you might expect, Python does not contain a fish() function, and so we get an error message saying, "NameError: name 'fish' is not defined." This shows another important principle for working with Python: You can freely try things out at any time without fear of breaking anything. If Python can't understand what you want to accomplish, or you haven't quite figured out how to do something, it will calmly respond with an error message and will not run any more code in that cell until you figure out what needs to be fixed. The error messages from Python are not always easy to decipher, but with some strategies that we will discuss in future chapters, you can break down the problem and figure out how to get the code to do what you want.

Finally, it's important to remember that Python is case sensitive. This means that my_Family_Ages is different from my_family_ages. If we misspell the name of the data object by messing up the capitalization, we will get an error because Python will interpret that code as a reference to a variable that does not exist.

Let's take stock for a moment. First, you should definitely try all of these lines of code noted on your own notebook. You can read about the lines of code in this book all you want, but you will learn a lot more if you actually try things out. Second, if you try a command that is shown in these pages, and it does not work, you should try to figure out why. Begin by checking your spelling and punctuation because Python is picky about how commands are typed. Remember that capitalization matters in. If you verify that you have typed a command as you see in the book, and it still does not work, try to go online and look for some help. There's lots of help at http://stackoverflow.com and at many other websites. If you can figure out what went wrong on your own, you will probably learn something valuable about working with Python. Third, you should take a moment to experiment with each new set of commands that you learn. For example, using only the methods discussed earlier in the chapter, you could do this totally new thing:

```
oldest = max(my_family_ages)

youngest = min(my_family_ages)

youngest, oldest
```

What would happen if you ran these three lines of code? What would you see? Think about how that worked, and try to imagine some other experiments that you could try. The more you experiment on your own, the more you will learn. Some of the best stuff ever invented for computers was the result of fiddling around to see what was possible.

SLICING LISTS

Python provides the capability to access individual elements within a list and calls this technique "slicing." For example, we can choose the first three elements of the my_family_ages vector:

```
my_family_ages[:3]

[43, 42, 12]
```

This technique is called "slicing" because we are slicing off one part of the list. It takes a while to get used to all of the features that slicing offers. For one thing, Python always starts counting things at zero, rather than one, like people do. So the first element of every list is actually element zero. Also, the colon character is used to set a range of values, and if you leave something out on either side of the colon, Python has a rule for what that shortcut means. In this case our expression inside the square brackets, ":3", is actually a shortcut for "0:3", which if we were to say it in words, would mean that we want to start at zero and go all the way up until before we get to three. As a result, the output shows the values 43, 42, and 12, which are the exact things we want if we wanted to use slicing to review the first three items in the list.

```
my_family_ages[-3:]

[12, 8, 5]
```

Python is also clever about using negative indices for slicing. A negative index counts backward from the end of the list. So the example in the code cell here starts three entries back from the end and then goes to the end. You may want to practice some additional examples to help get used to the different ways you can use slicing to access the elements of a list.

THE VIRTUAL MACHINE

When using Google Colab, Kaggle notebooks, or hosted notebooks provided by any other service, it is important to keep a few things in mind. First, as mentioned, when you first start running code in a notebook, it will take a moment to begin while the server initializes your virtual machine. In Colab you will know that process is complete when you see a small indicator for memory (RAM) and disk space on the upper right of your browser window. The virtual machine will usually keep running as long as you keep working, but if you leave your browser idle for a while, it will disconnect from the virtual machine. A dialog will come up on the screen asking if you want to reconnect, and if you do so, you can pick up right where you left off. All hosted notebooks have time limits, however, so ultimately your virtual machine will shut down, and everything you have stored in its memory will disappear. Usually, this is no problem because you can restart your notebook and run it from the beginning to get back to where you were. But if you are creating any data products that you want to reuse later, you should include a line of code that stores the data in a file. You can upload and download files to and from your virtual machine whenever it is running.

Additionally, you may have noticed from the examples that once a variable has had a value assigned to it, that variable persists and is visible in all of the later code cells in that notebook. This is usually a good thing: It is valuable to write your notebooks with small chunks of code in each cell rather than trying to cram everything into one cell. So, if a code cell runs (without error), and the code assigns values to a variable, like my_family_ages = [43, 42, 12, 8, 5], that variable and the list it contains will be visible in all subsequent code cells in this notebook. Now is as good a time as any to also urge you to include many comments in your code (any text following the # is a comment). Comments are a gift to your future self that will save you a lot of head-scratching when you come back later to look at your code.

SHARED PYTHON CODE LIBRARIES: THE PACKAGE INDEX

As an open-source language, Python is used by millions of people worldwide. Some of these people, when they have a problem to solve with code, will write some shareable software. After testing the software, they may choose to bundle it into a "library" or "package" of code (the two terms are often used interchangeably) for public release. The website https://pypi.org is the Python Package Index and it is the main repository for shareable Python code. At this writing there are more than 350,000 projects stored in the repository—way too many for us to get to know them all. Fortunately, there is only a small handful of packages that we can use to address the vast majority of the analytic problems that we may face as data scientists. Here's a short and necessarily incomplete list:

- numpy: This stands for "numeric Python" and contains all kinds of useful utilities for manipulating numbers. This package contains all of the basic statistical functions that most people need, such as mean() to calculate the mean, or average, as most people call it.

- pandas: A core package for data scientists, pandas implements the most useful data frame structure in the Python world. "Data frames" are an essential tool for storing and manipulating quantitative data.

- matplotlib: The name says it all. When you want to produce plots of numeric data, matplotlib provides both simplicity and flexibility.

- sklearn: This is short for Scikit-Learn, which provides all of the most common machine learning and statistical learning code needed for doing quantitative analysis.

Throughout the chapters that follow, we will repeatedly import code from these packages. The code for the book also ends up using about a dozen other important packages that we will introduce when we need them. For now, the only other thing you need to know is about how to import a package and make it possible for your code to access the powerful capabilities that each package supplies. There are two steps involved: downloading the code from the repository to your environment and making aspects of the imported code available for your code to use.

If you are using the Google Colab notebook environment, the first step is often accomplished for you in advance. The folks at Google anticipated many of the major packages you might need, such as numpy and matplotlib, and preloaded them into the virtual machine. In a few instances, future chapters will use a package not included in the Colab virtual machine, in which case you will need "pip install" to get the code you want. For example, if we wanted to load a library for creating advanced graphics, we would want to download the plotnine package from the repository with the following code:

```
!pip install plotnine
```

The exclamation point in front of the line of code sends this as a command to the operating system of the virtual machine. In Colab and most other hosted environments the pip package manager is already included on the system, and it knows what to do when you ask it to install the package. A command like that one will generate several lines of output and will usually conclude with a message saying that the requested package has been installed.

If you are using a Python development environment or Jupyter notebooks on your own computer, you will have to install most packages yourself—even commonly used ones like sklearn. Look online for instructions on how to run a terminal, and issue pip commands from the terminal command line.

The second step is to make the package that you downloaded available for use by your code. Three valuable Python keywords will help you get this job done: "import," "from," and "as." Here's a common example:

```
import numpy as np
```

This code statement tells Python that you are interested in running some of the functions provided by numpy. The as keyword lets you supply an abbreviation that you will use to refer to the package in question. After this line of code runs, you could perform basic statistical functions like this:

```
np.mean(my_family_ages)

22.0
```

You can also tell Python more specifically which functions you would like to run by indicating the package name and the specific package of interest:

```
from plotnine import qplot

qplot(x=my_family, y=my_family_ages, geom='col')
```

Notice that in the second line of code, we call qplot() directly (i.e., without anything like "np." in front of it) because the from/import statement made the qplot function known to Python. By the way, that call to qplot() produces this nice plot, shown here as Figure 1.1:

To summarize what we have discussed so far, you now know the following things about Python (and about data):

- You can run Python on your computer or (preferably) in a hosted Jupyter notebook environment such as Google Colab.
- You can type Python code into code cells.
- You can create a list of numbers or strings. The Python punctuation for a list is the square brackets. Items in a list are separated by commas. For a list of strings, each string is enclosed in quotes.
- A list is one of the most essential forms of data storage in Python, and the elements of a list can be accessed through slicing.

FIGURE 1.1 ■ Sample Plot Using Python

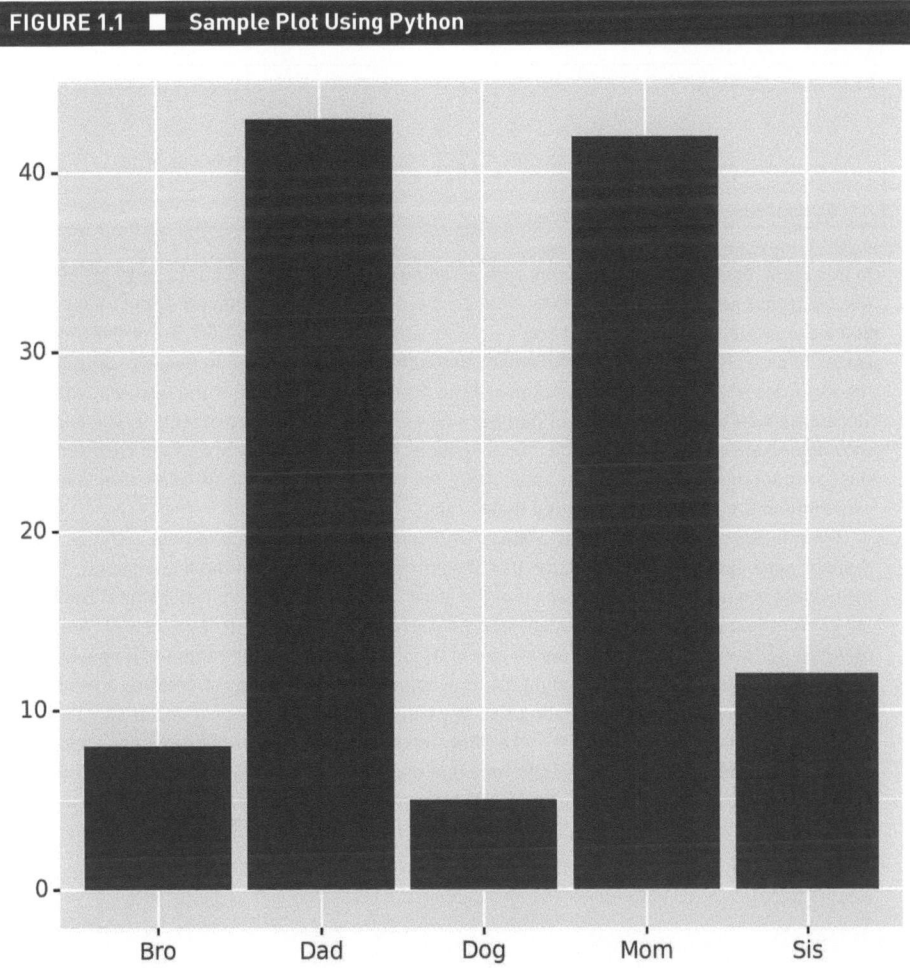

- A list can be stored in a named location using the equals sign. Once assigned, the variable is available for use in later code cells.

- You can get a report of the data object that is in any named location by typing that name by itself at the end of a code cell.

- Running a function, such as sum(), on a vector of numbers can transform them into something else. For example, sum() adds together all of the numbers in a list.

- Python is case sensitive. Keywords like "import" must be spelled exactly as they are shown. Variables you create, such as my_family_ages, have to use the exact same capitalization every time you reference them.

CASE STUDY: CALCULATING NPS

```
Case Key Points:
 - Define a list with a sample of likelihood to recommend
 - Calculate the number of promotors and detractors
 - Calculate NPS
```

In this book, each chapter concludes with a coding activity related to our case study of airline customer satisfaction. Let's take what we have learned in this chapter about using data and write some Python code that can calculate an overall Net Promoter Score (NPS) for a vector of data. The NPS is a measurement that some businesses use to quickly summarize the attitudes of a set of consumers toward the product or service that the business offers. To calculate NPS, we need a list of numbers where each number represents a *likelihood to recommend* answer to the question: "On a scale of 1 to 10, how likely are you to recommend this [product or service]." For our case study, we have asked flyers how likely they were to recommend the airline that provided their flight.

What follows is the code to calculate NPS. There are several ideas to note with this first realistic code example. First, note the use of comments (starting with the # character). Next, the newest and most interesting aspect of this code are the two list comprehensions. The first list comprehension contains this code fragment: [value for value in ltr if value > 8]. You can read this as "cycle through all of the values of ltr, one at a time, adding them to a new list as long as each value is larger than eight." A list comprehension is a way of creating a new and possibly different list from an existing list. We will be using list comprehensions quite frequently, so don't worry if it does not make total sense right now—you will have many opportunities to learn more about them. Finally, note that because the code says that a promoter must have an ltr score greater than eight and a detractor must have an ltr score smaller than seven, there are some values in ltr—specifically seven and eight—that will not make it onto either the promoter list or the detractor list! These survey respondents are considered to be in a middle ground where they are not necessarily dissatisfied with the product or service but are also not sufficiently excited about it to recommend it to a friend or family member.

```python
# Define a sample of LTR values
ltr = [9, 8, 3, 9, 7, 8, 9, 6, 7, 8, 9]

# What is the range (i.e., min and max) of the ltr list?
print("The range of LTR scores is:",
      min(ltr), "to", max(ltr))

# Create a new list with just the promoters
promoters = [value for value in ltr if value > 8]

# Calculate the proportion of the total LTR scores
proportion_promoters = len(promoters)/len(ltr)

# Create a new list with just the detractors
detractors = [value for value in ltr if value < 7]

# Calculate the proportion of the total LTR scores
proportion_detractors = len(detractors)/len(ltr)

# Calculate NPS based on the difference; make it a pct
nps = (proportion_promoters - proportion_detractors)*100

# Output NPS
print("The net promoter score is:", np.round(nps,1))

The range of LTR scores is: 3 to 9
The net promoter score is: 18.2
```

One important trick to making sense of a piece of code like this is to read all of the comments. When you are writing your own code, remember that the more comments you include, the easier it will be to understand later when you return to improve it or borrow parts of it to solve a new challenge. Above this code cell, we have also taken advantage one of the great features of a Jupyter notebook by including a text cell (sometimes also called a "text block"). This text cell contains a human-readable narration of what the code is trying to accomplish. We use these text cells in the code for every chapter to provide a more detailed explanation of the most interesting and challenging aspects of the code.

CHAPTER CHALLENGES

1. Use the append() method to add another family member's age onto the end of the myFamilyAges list like this: my_family_ages.append(17).

2. Use slicing to show only the first element of the my_family_ages list.

3. Use slicing with a negative index to show only the last element of the my_family_ages list.

4. Create a list comprehension using my_family_ages. The expression should show only those elements of the list where the person is older than 11.

5. Reverse the logic of the previous exercise by using a list comprehension to show only those family members younger than 12.

6. Assuming that you did Exercise 1, use the append() method to add another family member's name to the my_family list. Remember that a string must be enclosed in quotes. Then create a column graph with qplot() that shows the names and ages of all of the family members.

7. **Power User:** Calculate and display the mean and standard deviation of the ltr scores shown in the case study. Append another score to the ltr list that is equal to the mean of ltr. Recalculate and display the mean and standard deviation. Add a comment explaining what you observe.

2 ROWS AND COLUMNS

LEARNING OBJECTIVES

Explain what a DataFrame is and how data are organized in a DataFrame.

Create and use DataFrames in Python using pandas.

Access rows and columns in a DataFrame.

Manipulate and perform computations on a DataFrame.

Although we live in a three-dimensional world, where a box of cereal has height, width, and depth, it is a sad fact of modern life that pieces of paper, chalkboards, whiteboards, and computer screens are still only good at showing two dimensional displays. As a result, most of the statisticians, accountants, computer scientists, and engineers who work with lots of numbers tend to organize them in rows and columns. There's no good reason for this other than that it makes it easy to fill a rectangular area with numbers. Rows and columns can be organized any way that you want, but the most common way is to have the rows be cases or instances, and the columns be the attributes or variables. Take a look at this nice, two-dimensional representation of rows and columns found in Table 2.1:

TABLE 2.1 ■ Family Data

Name	Age	Hair	Weight
Dad	43	Black	188
Mom	42	Blonde	136
Sis	12	Brown	83
Bro	8	Brown	61
Dog	5	White	44

Pretty obvious what's going on, right? The top line, labelled in bold, is not part of the data. Instead, the top line contains the attribute or variable names. Note that computer scientists tend to call them attributes, whereas statisticians call them variables. Either term is OK, and both are used in professional data science settings. For example, age is an attribute that every living thing has, and you could count it in minutes, hours, days, months, years, or other units of time. Here we have the Age attribute calibrated in years. Technically speaking, the variable names in the top line are metadata, or what you could think of as data about data. Imagine how much more difficult it would be to understand what was going on in that table without the metadata. There are many different kinds of metadata: variable names are only one simple type of metadata.

If you ignore the top row, which contains the variable names, each of the remaining rows is an instance or a case. Again, computer scientists might call them "instances," and statisticians might call them "cases," but either term is fine. The important thing is that each row refers to an actual thing about which some data was recorded. In this case, all of our things are living creatures in a family. You could think of each of the items in the Name column as case labels, in that each of these labels refers to one and only one row in our data. Most of the time when you are working with a large dataset, there is a number used for the case label, and that number is unique for each case (i.e., the same number would never appear in more than one row). Computer scientists sometimes refer to a column of unique numbers as a "key." A key is useful, particularly for matching things from different data sources, and we will run into this idea again later. For now, though, take note that the Dad row can be distinguished from the Bro row, even though they are both Male. Even if we added an Uncle row that had the same Age, Gender, and Weight as Dad, we would still be able to tell the two rows apart because one would have the name "Dad" and the other would have the name "Uncle."

One other important note: Look at how each column contains the same kind of data all the way down. For example, the Age column is all numbers. There's nothing in the Age column like Old or Young. This is a valuable way of keeping things organized. After all, we could not run the mean() function on the Age column if it contained little pieces of text, like Old or Young. On a related note, every cell (i.e., an intersection of a row and a column, such as Sis's Age) contains only one piece of information. Although a spreadsheet or a word processing program might allow us to put more than one thing in a cell, a professional data handling program will not. Finally, see that every column has the same number of entries so that the whole forms a nice rectangle. When statisticians and other people who work with databases work with a dataset, they expect this rectangular arrangement.

CREATING PANDAS DATAFRAMES

Now let's figure out how to get these rows and columns into Python. One thing you will quickly learn about Python is that there is almost always more than one way to accomplish a goal. Sometimes the quickest or most efficient way is not the easiest to understand. In this case, we will build each column one by one and then join them together. This is somewhat labor intensive and not the usual way that we would work with a dataset, but it is easier to understand. First, we run this line of code to make the column of names:

```
my_family_names = ["Dad","Mom","Sis","Bro","Dog"]
```

One thing you might notice is that every name is placed within double quotes. This is how you signal to Python that you want it to treat something as a string of characters rather than the name of a storage location. If we had asked Python to use Dad instead of "Dad" it would have looked for a storage location (a data object or variable) named Dad. Another thing to notice is that the commas separating the different values are outside of the double quotes. If you were writing a regular sentence, this is not how things would look, but for computer programming the comma can only do its job of separating the different values if it is not included inside the quotes. Once you have typed this the line, remember that you can check the contents of my_family_names by typing it in a cell and then executing that cell:

```
my_family_names
['Dad', 'Mom', 'Sis', 'Bro', 'Dog']
```

Next, you can create a vector of the ages of the family members and then print out the elements in the list:

```
my_family_ages = [43, 42, 12, 8, 5]
my_family_ages
[43, 42, 12, 8, 5]
```

Note that to initialize the list, this is exactly the same command we used in the last chapter to define the list. If you had been using the same notebook, and you had kept Python running between then and now, you would not even have to retype this command because my_family_ages would still be there. But we suggest creating a new notebook for this chapter.

Now you have created lists named my_family_names and my_family_ages. If you look at the table earlier in the chapter, you should be able to figure out the code for creating my_family_hair and my_family_weights. In case you run into trouble, these commands also appear soon, but you should try to figure them out for yourself. After you type the command to create the new data object, you should also type the name of the data object to make sure that it looks the way it should. There are four variables, each with five values in it. Two of the variables are character data, and two of the variables are integer data.

Before we show you the Python code to create my_family_hair and my_family_weights, let's explore my_family_ages some more. We now know that my_family_ages is a variable, and that is a list; specifically it is a list of numbers. As we did in the last chapter, we can access each number individually using square brackets [and]. For example, if we want to output only the third element in my_family_ages, we could do the following:

```
my_family_ages[2]
12
```

Here are those two extra commands to define my_family_hair and my_family_weights in case you need them:

```
my_family_hair=\
    ["Black","Blonde","Brown","Brown","White"]
my_family_weights = [188,136,83,61,44]
```

Look out! We're starting to get commands that are long enough that they break onto more than one line. When this happens, you can add a "\" at the end of the line, and then Python knows that the code you are typing is not yet finished (that is what we did to define my_family_hair).

Now we are ready to tackle the DataFrame. In Python, a DataFrame is a specialized data structure, where each column is like a named list. Each column has the same number of entries, which is how we get our nice rectangular row and column setup, and generally each column also has its own unique name. To use DataFrames, we need to use a library known as pandas. If you are wondering why the name "pandas" was used, according to the Wikipedia page on pandas, the term "panda" comes from "panel data" (which is an econometrics term for multidimensional structured datasets). For any notebook that uses panda (and that will be most of our notebooks), you should do this import pandas command, typically at the start of the notebook:

```
import pandas as pd
```

Most Python programmers use pd as their preferred abbreviation for pandas. After this, when we want to call on pandas to do something for us, we will use the abbreviation followed by a dot, followed by the specialized pandas function that we want. For example, to create a DataFrame we can use the DataFrame function:

```
my_family=pd.DataFrame(
    {'names': my_family_names,
     'ages': my_family_ages,
     'hair': my_family_hair,
     'weights': my_family_weights
    })
```

The pd.DataFrame() function makes a DataFrame from the four lists that we provided. Notice that we have also used the assignment statement (i.e., the "=") to make a new stored location where Python puts the DataFrame. This new data object, called "my_family," refers to our newly created DataFrame. In a new code cell, type "my_family" and click Play to display what the DataFrame contains.

```
my_family

      names   ages    hair    weights
0     Dad     43      Black   188
1     Mom     42      Blonde  136
2     Sis     12      Brown   83
3     Bro     8       Brown   61
4     Dog     5       White   44
```

Notice how pandas has put row numbers in front of each row of our data. These are actual indices into the DataFrame. Later on, we can use row numbers like these to extract single rows or groups of rows from DataFrames.

Before we start exploring the DataFrame, let's look at another way we could have created the DataFrame. Using this method, we define a block of data and then the column names.

```
data = [ ['Dad', 43, 'Black', 188],
         ['Mom', 42, 'Blonde', 136],
         ['Sis', 12, 'Brown', 83],
         ['Bro', 8, 'Brown', 61],
         ['Dog', 5, 'White', 44]]

column_names = ['names', 'ages', 'hair', 'weights']

df = pd.DataFrame(data, columns=column_names)
```

EXPLORING DATAFRAMES

With a small dataset like this one, only five rows, it is pretty easy to take a look at all of the data. But when we get to a bigger dataset, this won't be practical. We need to have other ways of summarizing what we have. The first method reveals the data types that Python has used to store the information in the DataFrame. The ".dtypes" part of this line of code asks Python to reveal the dtypes (data types) attribute of my_family.

```
my_family.dtypes

names      Object
ages        int64
hair       object
weights     int64
dtype: object
```

In this case, we already knew the data types of my_family because we set that up in a previous command. In the future, however, we will run into many situations where we are not sure how Python has created a data object, so it is important to know how to use dtypes so that you can ask Python to report what kind of data has been stored. In this case, we can see that names and hair are objects (strings) and that ages and weights are integers. We can get additional information about the numeric variables using the describe function:

```
my_family.describe()

             ages        weights
count    5.000000       5.000000
mean    22.000000     102.400000
std     18.881208      59.070297
min      5.000000      44.000000
25%      8.000000      61.000000
50%     12.000000      83.000000
75%     42.000000     136.000000
max     43.000000     188.000000
```

We can see that the describe() method provides extensive information about both of the integer lists within the DataFrame. In the first line of output, we see the column names. In subsequent lines of output, we get seven different calculated quantities that help summarize each variable. There's no time like the present to start to learn about what these are, so here goes:

- count: shows the number of observations (we have confirmation that there are five observations/cases for each column).

- mean: the arithmetic average of all the values. For instance, the average age in the family is reported as 22.

- std: the standard deviation, which measures the dispersion of a dataset relative to its mean (i.e., how spread out the data is).

- min: refers to the minimum or lowest value among all the cases. For this DataFrame, five is the age of Dog, and it is the lowest age of all of the family members.

- 25%: the dividing line at the top of the first quartile. If we took all the cases and lined them up side by side in order of age (or weight), we could then divide the whole into four groups, where each group has the same number of observations. Like a number line, the smallest cases would be on the left with the largest on the right. The leftmost group, which contains one-quarter of all the cases, would start with five on the low end (Dog) and would have eight on the high end (Bro). So the first quartile is the value of age (or any other variable) that divides the first quarter of the cases from the other three quarters. Note that if we don't have a number of cases that divides evenly by four, the value is approximate.

- 50%: the median value of the case that splits the whole group in half, with half of the cases having higher values and half having lower values. If you think about it, the median is also the dividing line that separates the second quartile from the third quartile.

- 75%: the third quartile. This is the third and final dividing line that splits all the cases into four equal-sized parts. You might be wondering about these quartiles and what they are useful for. Statisticians like them because they give a quick sense of the shape of the distribution. Everyone has the experience of sorting and dividing things up—pieces of pizza, playing cards into suits, a box of cookies served on plates—and it is easy for most people to visualize four equal-sized groups.

- max: the maximum value. As you might expect, max displays the highest value among all of the available cases. For example, in this DataFrame Dad has the highest weight: 188.

That may seem like a lot of information about a small and simple dataset. Taking a step back, however, these statistics provide powerful ways of describing numeric data, and they are highly useful when you have a dataset with hundreds or thousands of cases. In particular, when we have a numeric variable, we almost always want to have at least two summaries of that set of numbers: a measure of central tendency and a measure of dispersion. The most common measures of central tendency are the mean, median, and mode. Common examples of measures of dispersion are the variance, standard deviation, and quartiles—each of these denotes how stretched or squeezed a distribution is. As you saw, in Python we can use the describe function to get the first two measures of central tendency (mean and median) as well as the standard deviation and the quartiles as measures of dispersion. We will explore standard deviation in more detail later in this book.

Although dtype and the describe function are both useful, sometimes we only want to preview a couple of rows in the DataFrame. Previously, we typed my_family and saw all the rows in the DataFrame. However, if the DataFrame has many rows, an alternative is to use the head() method. We supply an argument, in this case 2, to control the number of rows we would like to preview.

```
my_family.head(2)


     names  ages    hair  weights
0      Dad    43   Black      188
1      Mom    42  Blonde      136
```

You can see in the code that head() lists the first rows in the DataFrame. The tail() method is similar, but it lists the last rows in the DataFrame. The number of rows to display is the second argument to both head() and tail(). In our case, we used head() to show the first two rows of my_family DataFrame.

While using head() and tail(), we might also like to sort the DataFrame so that we can preview, for example, the youngest (or oldest) family members.

```
sDF = my_family.sort_values('ages', ascending=True)
sDF.head(3)


     names  ages   hair  weights
4      Dog     5  White       44
3      Bro     8  Brown       61
2      Sis    12  Brown       83
```

If you look at the output of our sorted DataFrame, which we called "sDF," you can see that the index of the row is not changed (e.g., the fifth row, with the row index of 4, still contains the record for Dog). We can override this and create new indices using the reset_index() method. This technique is demonstrated in the notebook file for this chapter. More generally, there are many additional ways to manipulate a DataFrame, and fortunately, the people who wrote pandas have provided excellent documentation, which you can view at the pandas site: https://pandas.pydata.org/docs/.

ACCESSING COLUMNS IN A DATAFRAME

Now that we can create and sort DataFrames, how can we get access to each of the variables stored there? We can use the name of the DataFrame together with the name of a column to refer to that column, like this:

```
my_family['ages']

0    43
1    42
2    12
3     8
4     5
Name: ages, dtype: int64
```

If you're alert, you might wonder why we went to the bother of putting all of this information in a pandas DataFrame and using this method of accessing a column when we could just use the original my_family_ages variable. This is an excellent question with an important answer. When we created the my_family DataFrame, we *copied* all of the information from the individual lists that we had before into a brand-new storage space. So now that we have created the my_family DataFrame, the ages column actually refers to a completely separate (but so far identical) vector of values. You can prove this to yourself easily by adding a new data point to the original vector, my_family_ages:

```
my_family_ages.append(11)
my_family_ages

[43, 42, 12, 8, 5, 11]
```

In the first line, we use the append() method to stick the value 11 onto the end of the original list of ages that we had stored in my_family_ages (perhaps we have adopted an older cat into the family). In the second line, we ask Python to report what the list my_family_ages now contains. We can see that Python reports that ages now contains the original five values and the new value of 11 on the end of the list.

Next, let's recheck the ages column in the DataFrame:

```
my_family['ages']

0    43
1    42
2    12
3     8
4     5
Name: ages, dtype: int64
```

When we ask Python to report the ages column in the DataFrame, you can see that we still have only the original list of five values. This shows that the DataFrame and its component column/list is now a completely independent data object. If we now wrote a line of code to change one of the original variables, such as my_family_names, that would have no effect on the information stored in the DataFrame.

By the way, if we do want to add some data to the end of a DataFrame, we need to provide a complete row of data so that the columns of the DataFrame all continue to contain the same number of observations. Here's an example using the loc() method to glue a new row onto the end of our existing my_family DataFrame:

```
new_row = \
{'names': 'Cat', 'ages':11, 'hair':'Yellow', 'weights':18}
my_family.loc[len(my_family), : ] = new_row
my_family
        names    ages     hair       weights
0       Dad      43.0     Black      188.0
1       Mom      42.0     Blonde     136.0
2       Sis      12.0     Brown      83.0
3       Bro      8.0      Brown      61.0
4       Dog      5.0      White      44.0
5       Cat      11.0     Yellow     18.0
```

In the first line of code (which actually breaks onto two line in the display), we create a small Python dictionary that contains the names of the columns and the values that we want for each cell. The column names need to exactly match what is in the my_family DataFrame, and the new values need to have the same data types as their respective columns. In the second line of code, we use the loc[] method, which provides a way of slicing the my_family DataFrame. In this case, by specifying the row as len(my_family) we are referencing the location a new row that will follow the existing final row of the my_family DataFrame (i.e., the row that currently contains Dog). The lonely colon after the comma says that we want to access all of the columns of my_family at the same time. You can see from the output that we have successfully added new_row, which has the data for Cat, onto the end of my_family.

For many different kinds of data in Python, including lists and DataFrames, square brackets allow for slicing subsets of the data. For example, my_list[3] would give us the fourth element of my_list. The use of square brackets in pandas is powerful but also complex, so let's look at a few examples. All of the following methods provide access to one particular cell in the DataFrame, namely, the row for Sis (the third row), which has the index of two, and the ages column, which can be referred to by name or by its number (it is the second column in the DataFrame and therefore has an index of one):

```
print(my_family['ages'][2])
print(my_family.ages[2])
print(my_family.iloc[2, 1])
print(my_family.loc[2, 'ages'])

12
12
12
12
```

The first line names the column and then provides a slicing index of 2 to access the Sis column. The second line refers to the ages column directly with a dot notation we have not used before. The third line uses "iloc," which means "integer location" and gives the index for row and then column. Finally, the fourth line uses "loc," which we saw in an earlier example—in this case we name the "ages" column to access its second cell. You should feel free to use whichever of these methods makes the most sense to you. It would be a good idea to practice at least one of these techniques by figuring out how to access a different cell in this DataFrame.

ACCESSING SPECIFIC ROWS AND COLUMNS IN A DATAFRAME

Keeping in mind that a DataFrame is a rectangular, two-dimensional structure, we can also access an individual row or column. Given our previous examples using loc and iloc, here's how we can use a shorthand for accessing a complete column of a DataFrame by leaving the row index empty:

```
my_family.loc[ : , 'ages']

0    43
1    42
2    12
3     8
4     5
5    11
Name: ages, dtype: int64
```

Likewise, a shorthand for taking a whole row of a DataFrame is to leave the column index empty. For example, we can look at the second row of my_family:

```
my_family.iloc[1, : ]
names        Mom
ages          42
hair      Blonde
weights      136
Name: 1, dtype: object
```

We can also supply a list of rows instead of only one row, like this:

```
my_family.iloc[1:3,:]

     names    ages    hair     weights
1    Mom      53      Blonde   136
2    Sis      23      Brown    83
```

Finally, we can subset both rows and columns at the same time:

```
my_family.loc[1:3,['ages', 'hair']]

        ages      hair
1        42       Blonde
2        12       Brown
3         8       Brown
```

If you look closely at the notation used to select the columns, you will see that 'ages' and 'hair' are enclosed in a list using square brackets. Providing a list of columns as the column selector can be a handy method of examining two or more pieces of data side by side to look for patterns. We can also take an expression like the one in this code cell and assign it to a new variable: That new variable will provide a "view" into the original DataFrame without copying the selected pieces of my_family into a new location.

GENERATING DATAFRAME SUBSETS WITH CONDITIONAL EVALUATIONS

When working with data, we often need to filter certain rows, columns, or cells that meet certain criteria. For example, we might want to examine or work with a subset of family members in a certain age range. This filtering strategy uses a logical condition to choose elements of a DataFrame. So rather than defining specific indices to select elements from the DataFrame, we can also tell Python, via a list of Trues and Falses, which elements to include or exclude. Let's take a look at some examples. First selecting some rows by providing a list of True and False:

```
my_family[ [False, False, True, True, False, False] ]

        names     ages     hair       weights
2        Sis       12      Brown        83
3        Bro        8      Brown        61
```

The following code produces exactly the same result but shows how we can store our filtering "mask" (our list of Trues and Falses) in a variable:

```
just_kids = [False, False, True, True, False, False]
my_family[ just_kids ]
```

We can extend this idea of using a list of conditionals to get a subset of a DataFrame based on a conditional test of the data itself by placing a conditional evaluation directly inside the square brackets. In the example that follows, we test which family members are older than 21.

The expression "ages > 21" applies that conditional test to every element of the ages column within the my_family DataFrame, and the expression produces a list of Trues and Falses that is exactly the same length as ages. First, let's show how the conditional evaluation works by itself:

```
my_family.ages > 21

0        True
1        True
2       False
3       False
4       False
5       False
Name: ages, dtype: bool
```

Next, we will place that conditional expression inside the square brackets to use it as a row selector:

```
my_family.names[my_family.ages > 21]

0       Dad
1       Mom
Name: names, dtype: object
```

Take note of the fact that we have used ages as a conditional selector variable, but the expression as a whole reports the names of the family members that match the condition. This shows that we can use a filter on one variable to affect what gets displayed for any other variable. Take the time to get a clear understanding of how this logic works when using conditional evaluations because it is a powerful feature that we will use over and over again as we transform and repair datasets.

We can also identify all the family members who do not meet a certain condition, and these tests can include evaluation of a character string variable. In the example that follows, note how in Python, the exclamation point provides a negation of a condition (i.e., it means "not" in the logical test):

```
my_family.names[my_family.hair != "Brown"]

0       Dad
1       Mom
4       Dog
5       Cat
Name: names, dtype: object
```

Finally, we can also construct an expression that performs multiple conditional tests at the same time, such as all the family members older than 6 but younger than 20:

```
my_family[ (my_family.ages > 6) &
           (my_family.ages < 20)]

     names    ages    hair    weights
2     Sis      12     Brown      83
3     Bro       8     Brown      61
5     Cat      11     Yellow     18
```

By the way, you might be wondering why we were able to have a piece of code that broke across two lines without having to use the line continuation character (backslash). In this case, Python reads the ampersand and knows that there must be another logical expression coming up, so it keeps reading the next line. You could add the backslash as the last character of the first line and get the same result (and no syntax error).

Finally, one way to make sure that your conditional works correctly, or to figure out a problem if it does not, is to copy only the conditional expression, run it by itself, and look at the resulting mask of True and False values that it produces.

A QUICK REVIEW

So what new skills and knowledge do we have at this point? Here are a few of the key points from this chapter:

- Statisticians, database experts, and others like to work with rectangular datasets where the rows are cases or instances and the columns are variables or attributes.

- In Python, one of the typical ways of storing these rectangular structures is in an object known as a DataFrame (using the pandas package). Technically speaking, a DataFrame is a list of columns where each column has the exact same number of elements as all the other columns (making a nice rectangle) and where a column contains only one type of data in a given column.

- The dtype and the describe() methods can be used to reveal the structure and contents of a DataFrame. Using dtype shows the structure of a data object, whereas describe() provides summaries of numeric variables.

- The head() and tail() methods can be used to reveal a few of the first or last rows in a DataFrame.

- Min and max are often used as abbreviations for minimum and maximum; these are the terms used for the highest and lowest values in a list of values.

- The mean is the same thing that most people think of as the average. The mean and the median are both measures of what statisticians call "central tendency." The median is the same thing as the 50th percentile.

Quartiles are a division of a sorted vector into four evenly sized groups. The first quartile contains the lowest-valued elements, for example, the lightest weights, whereas the fourth quartile contains the highest-valued items. Because there are four groups, there are three dividing lines that separate them. The term "first quartile" often refers to the dividing line to the left of the median that splits the lower two quarters. The third quartile is the same idea but to the right of the median and splitting the two higher quarters.

The variables/columns of the DataFrame can be accessed using bracket notation or dot notation to specify the name of the DataFrame along with the name of the variable/column. We can view rows, columns or a subset of both using logical column and row names and/or row and column indices.

CASE STUDY: CALCULATING NPS USING A DATAFRAME

Let's practice working with DataFrames by setting up a small number of survey responses, specifically, six surveys with ltr (likelihood to recommend) values of 9, 9, 7, 6, 8, 7 and the type of travel also defined as follows ("Business travel ", "Business travel ", "Business travel ", "Mileage tickets ", "Personal Travel ", "Personal Travel "). Given this, is there a difference in NPS, comparing all survey responses to just the business travel?

To do this analysis, we first need to create a DataFrame that represents the six surveys.

```
ltr = [9, 9, 7, 6, 8, 7]
type_of_travel = ["Business travel", "Business travel", "Business \
travel", "Mileage", "Personal Travel", "Personal Travel"]

survey = pd.DataFrame({'ltr': ltr,
                       'type_of_travel': type_of_travel})

survey.head()

    ltr     type_of_travel
0   9       Business    travel
1   9       Business    travel
2   7       Business    travel
3   6       Mileage
4   8       Personal    Travel
```

Next, we can calculate the number of promoters and detractors:

```
num_p = sum(survey.ltr > 8)
num_d = sum(survey.ltr < 7)
print('Number of promoters:', num_p)
print('Number of detractors:', num_d)


Number of promoters: 2
Number of detractors: 1
```

This use of the sum() function depends on a little trick: When we evaluate a logical expression, a True is represented by one, and a False is represented by zero. So we can easily count the number of True values simply by using the sum() function.

Now that we have a count of prompters stored in num_p and a count of detractors stored in num_d, we can use those values to calculate overall NPS:

```
total=len(survey)
nps = float(num_p - num_d)/total*100
nps


16.666666666666664
```

Next, we will use pandas to create a subset of the data where the type_of_travel is limited to business travel tickets only.

```
bus_travel_df=\
    survey[survey.type_of_travel=="Business travel"]

len(bus_travel_df)

3
```

The first two lines use the logical filtering that we demonstrated earlier in this chapter. The third line of code helps confirm that the resulting DataFrame contains only the three rows that match the logical condition. Next, we can calculate the NPS for only this subset of the data. We do not print out the intermediate counts of promoters and detractors, but you could easily add that code in a new code cell.

```
bus_num_p=sum(bus_travel_df.ltr>8)
bus_num_d = sum(bus_travel_df.ltr < 7)

bus_total = len(bus_travel_df)
bus_nps = float(bus_num_p - bus_num_d)/bus_total * 100
bus_nps

66.666666666666
```

As we can see from the output, in this dataset, people flying on business class tickets have a higher NPS when the comparison group is all passengers. This is a small dataset and therefore difficult to make generalizations from, so we would want to have more data before trying to draw any firm conclusions. In addition, it would be helpful to talk to subject matter experts (SMEs) to develop an understanding of why business class customers might include a greater proportion of promoters than the general population of air passengers. Of course, if you have taken a flight yourself, and seen how comfy the seats in business class are, perhaps you are already an SME!

CHAPTER CHALLENGES

1. Use the list notation (the square brackets) to create a new variable containing the favorite food of each family member. For example, your list could contain the entry "Pizza". Make sure that your new variable includes exactly five values. Call the new variable my_foods.

2. Add your new variable to the my_family DataFrame. If you were running the code while reading this chapter you will have my_family_names, my_family_ages, my_family_hair, and my_family_weights already available. Otherwise, you will need to type in the data for those variables as shown in this chapter.

3. Rerun the describe() function on my_family. Did the output change? Why or why not? How can you see the data type for the new column?

4. Create an expression that shows a list of True and False values based on the age of each family member. The variable should be True if my_family['ages'] is less than 40. In other words, your index will be True for kids and False for adults. Assign the results of your expression to a new variable called my_index.

5. Use my_index from the previous problem to show the favorite foods for each child in the family.

6. As was mentioned in this chapter, you use the exclamation point to negate a set of Boolean values by changing each True to False and vice versa. Adapt the expression from the previous problem to show the favorite foods for each adult in the family.

3 DATA MUNGING

LEARNING OBJECTIVES

Describe what data munging is.

Describe how to read a CSV data file.

Practice using the pandas Python library to select, filter, remove, and rename rows and columns.

Assess why data scientists need to be able to munge data.

Professional data scientists frequently face the task of preparing, cleaning, reshaping, and repairing a dataset before beginning the analysis process. In fact, experienced professionals know that these activities often take more time than the analysis itself. In 1960, a group of students at the Massachusetts Institute of Technology came up with a somewhat sarcastic term for these necessary but inconvenient preliminary operations: "munging." Munging was an acronym that originally meant "mash until no good." We, however, are going to mash our data until it is better to make sure that the conclusions we draw are as meaningful as possible and unaffected by errors, biases, or anomalies.

In this chapter we refer to data munging as the process of turning data with a bunch of junk in it into a nice clean dataset. Why is data munging required, and why is it important? There are a variety of forces of chaos at work on data that we collect, find, or receive: dates may be stored in the wrong format, extraneous material might appear at the beginning or end of a file, a data record might get cut off, a numeric field might be out of range, or a code to indicate missing data might be misinterpreted. Loads of trouble can crop up from the moment data is collected all the way up to the point that you are trying to run it through a statistical analysis.

Earlier in this book we explored simple datasets that we created inside of Python. Clearly, with larger datasets, it will not be practical to type data into a notebook. We need to be able to read in a large dataset and use tools in Python to clean it and get it ready to analyze. This kind of work may not be glamorous, but it is a big part of data science and important to get right. By the way, if you want a Data Munging coffee mug—it can be yours. Visit the website (http://www.cafepress.com/mf/17972553/i-love-munging_mugs). Money from the sale of mugs do not benefit the authors of this book—we just like the mug!

READING DATA FROM A CSV TEXT FILE

In this chapter, we will explore how to read in a dataset that is stored as a comma-delimited text file (known as a CSV file—which stands for comma separated values) that needs to be cleaned up. CSV files are inefficient to store but have the benefit of being human readable, making it easier to diagnose and repair problems. As we will see in future chapters, there are many formats that we might have to be able to process to get data into Python, but for now we will focus on this common file format. Our first real-world dataset will be U.S. census data. The U.S. Census Bureau has stored population data in many locations on its website, with many interesting datasets to explore. We will use one of the state-by-state population datasets available from among the files listed on this page: www2.census.gov/programs-surveys/popest/tables/2010-2011/state/totals/.

If you want to preview the file we will be using, point your web browser at that page, and click on the link for nst-est2011-01.csv; your browser will either download a CSV file or show a bunch of text information on your screen with the first few lines looking like this:

```
table with row headers in column A and column headers in rows 3 through
4. (leading dots indicate sub-parts),,,,,,,,, "Table 1. Annual
Estimates of the Population for the United States, Regions, States, and
Puerto Rico: April 1, 2010 to July 1, 2011",,,,,,,,, Geographic
Area,"April 1, 2010",,Population Estimates (as of July 1),,,,,,
,Census,Estimates Base,2010,2011,,,,, United
States,"308,745,538","308,745,538","309,330,219","311,591,917",,,,,
Northeast,"55,317,240","55,317,244","55,366,108","55,521,598",,,,,
Midwest,"66,927,001","66,926,987","66,976,458","67,158,835",,,,,
South,"114,555,744","114,555,757","114,857,529","116,046,736",,,,,
West,"71,945,553","71,945,550","72,130,124","72,864,748",,,,,
.Alabama,"4,779,736","4,779,735","4,785,401","4,802,740",,,,,
```

Although it is always a great idea to preview a dataset, showing the data in the browser will not help us transform or analyze it, so let's create some Python code that can read this dataset directly from the census website into our Python notebook. The first line of code makes the pandas library available. Don't forget that when we say "as pd," we are setting up an abbreviation for the library that we can use in later code, like we do in the last line of this block:

```
import pandas as pd

url = ('http://www2.census.gov/programs-surveys'
       '/popest/tables/2010-2011/state/totals/'
       'nst-est2011-01.csv')

census_df = pd.read_csv(url)
```

As discussed in the previous chapter, the pandas package provides the best way to use DataFrames in Python. In this situation, we are using a pandas function called read_csv() to read a CSV file directly from the web. The next three lines specify the location on the web of the file to load. The acronym URL refers to a uniform resource locator, also known as a web address. Note that with any long string, it might be easier for humans to read if we create the string on multiple lines and use the parentheses to combine the substrings. The URL must end up as one continuous string with no spaces or line breaks at the start or end of each substring. The final line of code reads the file. Let's take a look at what we get back. We can review the shape attribute to explore the structure of census_df:

```
census_df.shape

(66, 10)
```

The overall structure is 66 instances of 10 attributes, suggesting that the original census spreadsheet contains 66 rows and 10 columns of data.

When we examine the first column name after calling census_df.head(5), here's what we see:

```
table with row headers in column A and column head-
ers in rows 3 through 4. (leading dots indicate sub-parts
```

This is almost certainly not meant to be the name of a variable. It seems that read_csv() treated the first text it saw in the file as a variable label. It was flummoxed by the fact that this text was a note to human users of the spreadsheet. The real variable labels seem to be a few rows down. We might be able to at least partially fix that problem by asking read_csv() to skip a couple of rows before it starts to put data in the DataFrame. In the next block, we redo the previous command, but this time we specify "skiprows=2" as an argument to read_csv():

```
census_df = pd.read_csv(url, skiprows=2)
census_df.shape


(64, 10)
```

Now when we call census_df.head(7), we can see that the first column name is "Geographic Area" and that at least a few of the other column names seem to make more sense. Skipping a couple of rows therefore seems to have improved our input process. We can also use the dtypes function to explore the structure of census_df:

```
census_df.dtypes



Geographic Area                        object
April 1, 2010                          object
Unnamed: 2                             object
Population Estimates (as of July 1     object
Unnamed: 4                             object
Unnamed: 5                             float64
Unnamed: 6                             float64
Unnamed: 7                             float64
Unnamed: 8                             float64
Unnamed: 9                             float64
dtype: object
```

The output shows that our data types are a strange mix of strings (objects) and numbers (floats). Except for the first column, which labels the geographic area, we want everything to be numbers (more specifical, integers). Eventually we will have to fix these data types as well, but before we get to that, we should filter out the rows and columns that we don't need.

REMOVING ROWS AND COLUMNS

If you are following along in the Python notebook, you may have noticed that the first few rows contain population summaries for the United States as a whole as well as for several large geographic regions of the country. If we are planning to analyze individual population units, such as states and territories, we will have to remove these summary rows. Also visible in the data previews are extra columns that have been labelled "Unnamed: 4" through "Unnamed: 9." You can tell that these columns are not used because all of the values show as "NaN," which stands for "not a number." Our next data munging step should therefore be to get rid of unneeded rows and columns.

Before we get started, let's review how to access the elements of a DataFrame. As mentioned briefly in the previous chapter, Python's square brackets notation supports indexing into a list or DataFrame. For example, my_list[3] would give us the fourth element of my_list (remember that Python always starts counting at zero). Keeping in mind that a DataFrame is a rectangular, two-dimensional structure, we can address any element of a DataFrame using "iloc" (integer location) with a row and column designator: my_frame.iloc[0, 1] refers to the topmost row (i.e., row zero) and the second column. A shorthand for taking one whole column of a DataFrame would be to leave the row index empty: my_frame.iloc[:, 6] would give the seventh column of every row. Likewise, a shorthand for taking one whole row of a DataFrame is to leave the column index empty: my_frame.iloc[9,:] would give the entirety (i.e., every column value) of the tenth row. We can also supply a list of rows instead of only one row, like this: my_frame.iloc[[1,3,5],:] would return rows 2, 4, and 6 and would include the entirety of each row because we left the column index blank.

Using this knowledge, we can use the iloc[] slicer to get rid of stuff we don't need. We identified six header rows that we can eliminate like this:

```
census_df = census_df.iloc[6:, :]

len(census_df)

58
```

The row selector "6:" says that we want to get a subset of the original data that starts at row six (thereby effectively ignoring rows zero through five). We also leave the column selector empty so that we keep all columns for now. We assign the result back to the same data object, thereby filtering the original down to our new, smaller, cleaner version. To verify it worked, we check the length of the DataFrame (we could also use the shape attribute). The result shows that we have eliminated six rows.

Next, we can also see that of the 10 variables we got from read_csv(), only the first five are useful to us. We can use the following code to retain only the first five columns of the DataFrame:

```
census_df = census_df.iloc[:, :5]

census_df.shape

(58, 5)
```

RENAMING ROWS AND COLUMNS

Now we are ready to perform a few additional modifications. Although pandas does permit a wide variety of variable names—including ones that contain spaces—most analysts like to work with shorter variable names that contain no spaces. An underscore is sometimes used in place of a space to make a variable name easier to read. Let's set up some compact, readable column names, as shown here:

```
census_df.columns = ['state_name', 'april10census',
                     'april10base', 'july10pop', 'july11pop']
```

As the code shows, a pandas DataFrame has an attribute called "columns" that stores the names of the columns. We can overwrite the existing names by providing a list of the new names. Next, we will explore to the end of the DataFrame, as we did with the start of the DataFrame. We can use the tail() function to show us the last eight rows.

```
census_df.state_name.tail(8)

56                                  .Wyoming
57                                       NaN
58                               Puerto Rico
59    Note: The April 1, 2010 Population Estimates b...
60                        Suggested Citation:
61    Table 1. Annual Estimates of the Population fo...
62    Source: U.S. Census Bureau, Population Division
63                  Release Date: December 2011
Name: state_name, dtype: object
```

Take note of the expression we used to invoke the tail() method. By inserting state_name between census_df and tail(), we get a report that contains data only from that one column. We can see that the last few rows contain some Census Bureau notes that we don't need. Let's eliminate those rows like this:

```
census_df.drop([57], inplace=True)

census_df = census_df.iloc[:-5,:]

census_df.tail(5)
      state_name      april10census   april10base   july10pop    july11pop
53    .Washington     6,724,540       6,724,540     6,742,950    6,830,038
54    .WestVirginia   1,852,994       1,852,996     1,854,368    1,855,364
55    .Wisconsin      5,686,986       5,686,986     5,691,659    5,711,767
56    .Wyoming          563,626         563,626       564,554      568,158
58    Puerto Rico     3,725,789       3,725,789     3,721,978    3,706,690
```

The first line of code eliminates the empty row that has the index 57. Then the second line of code filters out the last five rows. Now the tail() method confirms that we have eliminated all of the end matter.

If you have a sharp eye, you may notice that Wyoming has the index number 56. That seems too large, given that there are 50 states plus Washington DC and Puerto Rico. The problem is that when we cut off the first eight rows in earlier code, pandas retained the original row indices. Although that is sometimes a useful behavior, in this case let's reset the indices so our rows are numbered starting with zero.

```
census_df = census_df.reset_index(drop=True)
census_df.head()

     state_name      april10census     april10base
0    .Alabama        4,779,736         4,779,735
1    .Alaska         710,231           710,231
2    .Arizona        6,392,017         6,392,013
3    . Arkansas      2,915,918         2,915,921
4    .California     37,253,956        37,253,956
```

We used the drop=True argument so that pandas would not create a new variable containing the old row index. The result is a DataFrame with 52 rows and five observations/columns. The rows are the 50 states plus the District of Columbia (aka Washington, DC) and the Commonwealth of Puerto Rico, a U.S. territory in the Caribbean Sea.

CLEANING UP THE ELEMENTS

Next, will change formats and data types as needed. In the first step, we will remove the dots from in front of the state names. Although this might seem only like a way to make the labels more visually appealing, keep in mind that the state names might be used in the future to link these population data to other information. For that linkage to work properly, we would need all of the state names to be spelled correctly and without any extraneous punctuation. For this step we will use a list comprehension, one of the powerful techniques that Python provides for looping through a collection of data:

```
census_df['state_name'] = \
     [element.replace('.', '') for
         element in census_df['state_name'] ]

census_df['state_name'][0:4]

0 Alabama
1 Alaska
2 Arizona
3 Arkansas
Name: state_name, dtype: object
```

How do you know that this is a list comprehension? When we see the "for" keyword inside an expression that has square brackets, that means the expression is a list comprehension. The second and third lines in the code here accomplish the main work we are trying to do. Consider the part of the expression beginning with "for" first: What it says is to loop through every cell in the census_df['state_name'] column. For each new cell we assign the value that we find into a temporary variable called "element." Next, take a look at the expression that comes before the "for" keyword: Here we use element.replace('.', ' ') to locate the dot character in the state name and replace it with the empty string (the part of that expression after the comma is two single quotes). Given the 50 states plus DC and Puerto Rico in the original column of data, we check 52 separate strings for those dots, returning a list of 52 strings without dots. In the last step of this list comprehension (which is actually the first line of code in this box), we assign the resulting list of 52 strings back to the state_name column.

Next, by looking at the data types of each of the columns, we can see that the populations are not numbers but rather objects (strings):

```
census_df.dtypes

state_name      Object
april10census   Object
april10base     Object
july10pop       Object
july11pop       object
dtype: object
```

In looking back at the values in the columns, we need to convert the data contained in the population columns to usable numbers. Remember that those columns are now represented as character data, and what we are doing is taking out the commas and making the digits that remain into numeric data:

```
census_df['april10census'] = \
    [int(element.replace(',', '')) for \
        element in census_df['april10census'] ]

census_df['april10base'] = \
    [int(element.replace(',', '')) for \
        element in census_df['april10base'] ]

census_df['july10pop'] = \
    [int(element.replace(',', '')) for \
        element in census_df['july10pop'] ]

census_df['july11pop'] = \
    [int(element.replace(',', '')) for \
        element in census_df['july11pop'] ]
```

```
census_df.dtypes
state_name        object
april10census      int64
april10base        int64
july10pop          int64
july11pop          int64
```

Although this might look like a lot of code, we are simply doing the same thing four times, each time to a different column. In each of the population columns, we accomplish two steps within each list comprehension: We replace each comma with the empty string, similar to what we did to remove the dot from the state name. Then, in addition to that, we are running the int() function to change the data type from string to integer. Finally, we confirmed that the new columns on the DataFrame were numeric by looking at dtypes. Let's crosscheck our results by looking at a small slice of the DataFrame:

```
census_df.head(4).iloc[ : , 0:2]
     state_name    april10census
0    Alabama          4779736
1    Alaska            710231
2    Arizona          6392017
3    Arkansas         2915918
```

In this line of code, we combine the head() method, which extracts the first few rows of the DataFrame, with the iloc[] slicer, which allows us to specify any rectangular subset of a DataFrame. In the result, we see four rows but only the first two columns of the data. This confirms that the population values look like well-formed numeric values. Notice how much effort we have put in to conditioning the data we got to make it usable for later analysis. An important, and sometimes time-consuming, aspect of what data scientists do is to make sure that data are fit for the purpose to which they are going to be used.

SORTING AND GROUPING DATAFRAMES

Now that we have a real dataset, let's do something with it. Demographers are scientists who examine the causes and consequences of population changes. One key value they examine is how much a population changes over time. Let's calculate the change in population for each state. First, we make a copy of our data to ensure that we do not mess up the original:

```
pops_df = census_df.copy(deep=True)
```

We have supplied an argument to this call to the copy() method that helps ensure that we do not change the original data. By using "deep=True" we create a deep copy of census df and place that copy in pops_df. This severs any connection between the two datasets. If we had used "deep=False," then pops_df would be nothing more than a different way of viewing the original census_df data. Next, we will calculate the changes in population for each state and add the resulting list as a new column in our DataFrame:

```
pops_df['increase'] = \
   (pops_df['july11pop'] - pops_df['july10pop']) / \
   pops_df['july11pop']
pops_df.head(4).loc[ : , ['state_name', 'increase'] ]

       state_name     increase
0      Alabama        0.003610
1      Alaska         0.011861
2      Arizona        0.010698
3      Arkansas       0.005579
```

The first three lines calculated the difference between the 2011 population and 2010 populations (both measured in July of the respective years) and expressed the result as a proportion of the 2011 population. The last line once again uses the head() method to show the first three rows, and here we have used the loc[] slicer to refer to the two columns that we want to see. In a different browser window, we verified the resulting value for the first row (for Alabama) with a calculator—always a valuable cross-checking step if you are depending upon the result for later operations.

Our new pops_df DataFrame contains a calculation of the population increase for each state. By sorting these data we can call attention to the state with the largest increase and the state with the largest decrease in population. In this next code block, we create sorted_pops, which is a view into the existing DataFrame but with the rows sorted in decreasing order based on our new 'increase' column:

```
sorted_pops = \
pops_df.sort_values(by=['increase'], ascending=False)

print("Largest increase:")
print(sorted_pops.head(1).loc[:,['state_name','increase'] ])

print("\nLargest decrease:")
print(sorted_pops.tail(1).loc[:,['state_name','increase'] ])

Largest increase:
           state_name   increase
8    District of Columbia   0.021172

Largest decrease:
     state_name    increase
51   Puerto Rico   -0.004124
```

One interesting thing about this result is that neither the first row, District of Columbia, or the last row, Puerto Rico, is a state! District of Columbia showed a 1-year increase of more than 2 percent. In the same year, Puerto Rico lost 0.4 percent of its population. We know that this was a decrease by the minus sign shown for Puerto Rico's value. If you go back to the original calculation of the 'increase' column, you will find that if the 2011 population is smaller than the 2010 population, that will cause the result of the calculation to be negative.

GROUPING WITHIN DATAFRAMES

We will close this data exploration with one additional step. When communicating results of a data analysis, it is often easier for audiences to understand what is happening when you can display results for two contrasting groups. Sometimes these contrasting groups occur naturally in the world: for example, we could compare dogs and cats on variables such as weight, diet, or daily sleep time. In other situations, we can create our own groupings to illustrate some contrast that we want to make. For our state population dataset, we are going to split the data into two equal groups: all of the largest states in one group and all of the smallest states in the other group. Here are some lines of code that accomplish the split:

```
median_pop = sorted_pops['july11pop'].median()

sorted_pops['above_median'] = \
[val > median_pop for val in sorted_pops['july11pop']]

grouped_pops = sorted_pops.groupby('above_median')
median_pop

4120607.5
```

That first line of code calculates the median population value for July 2011 and stores it in a new variable called median_pop. You may remember from our earlier discussion of statistics that the median is the 50th percentile, and thus it is the value of a variable where half of the values in the dataset are larger and half are smaller than that value. For our dataset, a value of 4,120,697.5 divided our 52 cases into two groups of 26. The second line of code produces a new grouping variable that helps us accomplish the split. Here we use a list comprehension to create a "Boolean" variable—a variable that can either be True or False. The code computes a True for each case where the July 2011 population is above the median and a False for each case where it is below the median. We store the result in a new column on our sorted_pops DataFrame. Finally, in the last line of code we use the groupby() method to split the DataFrame into two separate pieces.

Now we are ready to look at a couple of contrasts. We have called our new grouped DataFrame "grouped_pops," and every time we ask a question to that new DataFrame, we will get two answers: one for the below-median group and one for the above-median group. In this next line of code, we ask for a report of the maximum population increase value for each of our two groups:

```
grouped_pops.increase.max()

above_median
False    0.021172
True     0.016406
Name: increase, dtype: float64
```

What the first line of output shows is that for the smaller states and territories, the maximum increase in population among states or territories is about 2.1 percent. From earlier results we know exactly which area we're talking about: the District of Columbia. The second line of output shows that among the largest states and territories, however, the maximum population increase is only 1.6%. The state in question is Texas, which we have verified with an additional line of code in the notebook. This result makes sense: With a population of more than 20 million, it is much more difficult for Texas to grow by 1 or 2 percent than it is for Washington, DC.

Out of curiosity, we can repeat the analysis but this time asking for a report of which state or territories had the minimum population increase (i.e., the greatest decrease) within each of our two groups:

```
grouped_pops.increase.min()

above_median
False   -0.004124
True    -0.000097
Name: increase, dtype: float64
```

Again, these results produce a familiar result in the first line. The largest decrease in population among smaller states and territories is about –0.004 (–0.4 percent). We know from earlier results that this value is from Puerto Rico. The contrasting result for larger states, –0.000097, turns out to be Michigan, which lost about 1,000 people out of a population of more than 9.8 million. That's a pretty small number of people to lose, especially for a northern state: Some northern states have been steadily losing population for the last century.

These results begin to show us the power of doing data analysis with pandas DataFrames. These rectangular structures offer many built-in methods that allow us to slice and dice the data in ways that can reveal important trends and patterns. Looking back, here are the new skills we developed in this chapter:

"Data munging" is our term for the processes we need to do to get data in shape and ready to analyze. These processes are also sometimes called "data screening" or "data cleaning."

The pandas package gives us the power to read in CSV datasets either directly from a file or from a location on the web.

We must carefully review variable names and data types to make sure that the format, structure, and types of data fit our expectations for the columns and rows we expect to find. The essence of data munging is finding and fixing problems with the configuration of data.

Pandas provides the ability to specify subsets of rows and columns that we can use to discard empty or erroneous elements from a dataset. The loc[] and iloc[] slicers provide two important ways of doing this.

List comprehensions provide a way to loop through the elements of a column to perform transformations on each element/cell. In this chapter we did string substitutions and changed the data type of columns using list comprehensions.

Pandas provides sorting and grouping capabilities that help us organize the data. Data that have been sorted make it easy to locate the highest and lowest values. Grouped data makes it easy to compare contrasting subsets of the data.

CASE STUDY: READING, CLEANING, AND EXPLORING A SURVEY DATASET

```
Case Key Points:
 - Read a CSV file that has survey data
 - Clean the dataset (survey) by removing NAs
 - Explore LTR (likelihood to recommend) calculating the mean of LTR
   across different groups within the survey
```

Let's explore a small airline survey dataset. First, we need to read in the small airline survey CSV datafile. Next, after we have the survey dataset, we will explore it by first calculating the overall mean likelihood to recommend (noted as LTR in this dataset). The code to begin this analysis follows:

```
small_survey = pd.read_csv ("https://"
                            "raw.githubusercontent.com/"
                            "jmstanto/IDSpython/main/"
                            "smallSurvey.csv")

small_survey.dtypes
```

```
Type.of.Travel                  Object
Airline.Status                  Object
Gender                          Object
Age                              int64
LTR                            float64
dtype: object
```

We have previously used dtypes to examine DataFrames. We can see that there are five attributes (columns) where LTR and age are numbers and the others are strings. Next, let's look at the first few rows in the dataset. Although we could have easily looked at all the columns at once, next we create a subset of columns, mainly to practice for when our DataFrames have too many columns to display in one chunk of output:

```
small_survey[
  ['Age', 'Type.of.Travel', 'LTR'] ].head()

    Age   Type.of.Travel    LTR
0    22   NaN               7.0
2    39   Personal Travel   8.0
3    64   Personal Travel   6.0
4    26   Business travel   9.0
5    50   Business travel   7.0
```

In looking at the output, we note one issue with the dataset. Specifically, there is an NaN for Type.of.Travel in the first row. Remember that in Python, NaN stands for "not a number," so it is Python's way of signaling that there is a missing data element. To clean this dataset, we will remove any row of data without an LTR response because, without an LTR response, that row will not be helpful in understanding drivers for improving LTR. For now, however, we can leave NaNs in the other columns (e.g., if someone doesn't provide their age) unless and until they cause a problem in an analysis. Most analysts would consider this as a "conservative" strategy: don't discard rows of data that contain missing fields unless these missing fields are interfering with an analysis. Pandas provides the dropna() method for eliminating rows with missing data. In the middle line of code that follows, we direct pandas to drop a row only if there is a missing value on LTR:

```
print("# of cases before dropping NaN:", len(small_survey))
small_survey = small_survey.dropna(subset=['LTR'])
print("# of cases after dropping NaN:", len(small_survey))

# of cases before dropping NaN: 500
# of cases after dropping NaN: 499
```

The output shows that one case was lost due to a missing data field in LTR. Next, for safety we will make a copy of the small_survey dataset. In addition, because of the aggregation operation we are about to do, we subset the data to include only the numeric variables Age and LTR:

```
numeric_survey = small_survey.loc[ : ,['Age', 'LTR']]

numeric_survey.columns

Index(['Age', 'LTR'], dtype='object')
```

The output confirms that we have saved only the Age and LTR columns in this new version of the dataset. In this final step, we are going to ask pandas to accomplish two steps. In the first step, we will use the groupby() method that we demonstrated earlier in the chapter to organize the data. We will create a new variable to do this grouping that converts the Age variable into decades by using rounding. You probably have experience rounding a number by eliminating one or more decimal places. For example, if I want to round 10.2 down to no decimal places, I will get 10. Interestingly, there is nothing stopping us from rounding even further. The "–1" in this rounding command rounds off the ones place of each value of Age:

```
numeric_survey['rounded_age'] = \
    round(numeric_survey['Age'],-1)
print("Min:", min(numeric_survey['rounded_age']))
print("Max:", max(numeric_survey['rounded_age']))

Min: 20
Max: 80
```

The two print() statements produce output that suggests that we have rounded each value of Age to the nearest 10. We should now have about seven categories of Age, each covering a 10-year period. Now, in the final analysis step, we can group the data based on these age decades and produce an aggregation report. The aggregation report gives an overall picture of what each group contains:

```
numeric_survey.groupby('rounded_age').agg(
  ['mean','count'])
              Age mean    count    LTR mean    count
rounded_age
20            19.017544   57       6.754386    57
30            30.215190   79       8.063291    79
40            39.841667   120      7.825000    120
```

```
50         49.741176      85      8.117647    85
60         59.518987      79      6.670886    79
70         69.944444      36      6.000000    36
80         79.418605      43      5.674419    43
```

The results table shown in this output contains a lot of useful information. On the far left you see a column that shows the seven age decades. Remember that this is our grouping variable, so we are going to get some output for each of these groups. Next, there are two pairs of columns, one pair for Age (the nonrounded version of this variable) and one pair for LTR. For each variable we can see the mean value for that age group as well as the count of the number of rows that went into calculating that mean.

Pay close attention to the count columns: Notice that the counts are identical for Age and LTR. This is a good sign that there are no problematic missing data patterns. In fact, because we already used the dropna() method earlier to eliminate any rows with missing LTR, we can be assured by looking at these counts that there are no missing entries for Age either.

By grouping the ages, and then calculating and displaying the mean LTR, we can see a definite trend where those greater than 70, on average, have a lower likelihood to recommend. Although there clearly is more analysis that would need to be done before understanding the drivers of higher and lower NPS, this high-level analysis starts to provide some clues that could be explored.

CHAPTER CHALLENGES

1. Use the seaborn library to import a built-in dataset. First, import the seaborn library, and then load an example dataset with the following line of code:

   ```
   tips=seaborn.load_dataset('tips')
   ```

   How many rows and columns does this tips dataset have?

2. Use dtypes to reveal the names and data types of the columns in the tips dataset.

3. Sticking with tips, use the square brackets notation to select the second observation in the second column.

4. Continuing to work with tips, select the size column. Run the head() command on this column.

5. Create a new column, which is the tip percentage (i.e., tip / total_bill)

6. Group the tips dataset by day, and output the average tip by the day.

7. Group the tips dataset by day, and output the average tip by the day as well as the average bill by day.

8. Sort the groups by increasing average tip (grouped by day).

9. Go to the LendingClub website (https://figshare.com/articles/dataset/Lending_Club/22121477), download a CSV file, and read in the file using read_csv(). Then, clean up the dataset, making sure all the columns have useful information. This means you must explore the dataset to understand what needs to be done! One trick to get you started is that you might need to skip one or more lines (before the header line in the CSV file). There is a skip parameter that you can use in your read_csv() command.

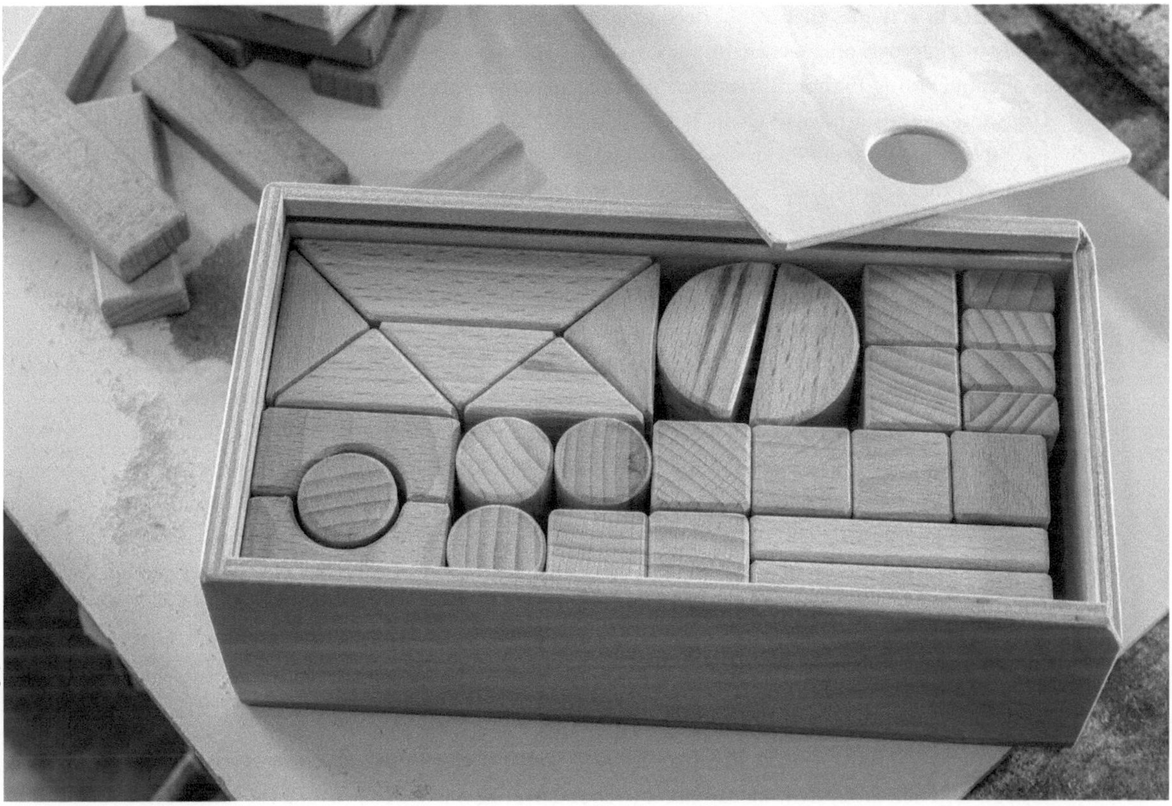

iStock.com/Sergey05

4 WHAT'S MY FUNCTION?

LEARNING OBJECTIVES

Explain the benefits of writing and using functions and methods.

Write a basic function in Python.

Create a custom class in Python that includes a method.

Early in the evolution of programming languages, researchers realized that there were certain sequences of activities that would need to occur repeatedly. Rather than typing the same set of instructions over and over again, they decided to create reusable units of code. In many languages, including Python, these reusable units are called "functions." Other programming languages may also refer to them as procedures or subroutines, although these terms may have somewhat different meanings in other languages. We have already been using functions, such as min() and max(), that are provided with Python. Those functions were written by the creators of Python, and they are provided by default when Python is installed, meaning that we don't have to show Python how to compute the minimum or maximum with detailed code—we can simply ask these functions to perform the necessary work instead. Because these functions are already written and are known to Python, we can simply call them in our Python code whenever we need them without worrying about how they work on the inside.

People who use Python regularly often create their own new functions so that they can achieve that same level of convenience by automating a new process or calculation. A collection of functions can also be published as a Python library if the authors expect that many other people will want to reuse it. We have publicly available libraries such as pandas because a group of people decided that functions for manipulating rectangular datasets would be helpful to a lot of other folks. Even if we don't intend to publish a new library ourselves, however, creating our own functions can help keep our code readable and bug free.

WHY CREATE AND USE FUNCTIONS?

There two important reasons that people create and use functions. The first is reusability. Once a function is defined, we can use it over and over again. We can invoke the same function many times in our program, and this saves us work. In addition, once a function is tested, and we believe it works correctly, we can also expect it to work correctly in the future, which can save time on debugging.

The second reason is a concept from software engineering known as "abstraction." Abstraction means that after a function has been created, we don't need to remember the details of how it works to use it. This can save a lot of brainpower because we can more easily focus on higher-level concepts once the low-level details are handled. Once a function has been created and is known to Python, we only need to know the following four things when we want to use a function:

1. the name of the function,
2. its purpose (i.e., why and when we would want to use it),
3. what arguments (inputs) must be supplied to the function, and
4. what kind of result (output) the function returns.

We have already used arguments a little while ago with functions such as min() and max(). For these two functions, we supplied a list of numbers so that they could do the necessary calculations.

We did not discuss the idea of "arguments" at that time—we just used the function. Argument (also known as "parameter") is a term used by computer scientists to refer to some information that is sent into a function to help it do its job. For min(), we "passed in" one argument, the list of numeric values among which we should find the smallest number. Other functions expect more than one argument. One such function is read_csv(), which can receive additional arguments, such as how many rows to skip when reading a CSV file to create a DataFrame. Sometimes these additional arguments to read_csv() can be made optional, which means that function only needs it in certain circumstances and/or can supply a default value if needed.

Functions also often have outputs, or what are sometimes called return values. Typically, a return statement will be the last line of code in a function and it will send back one or more pieces of data to whoever called it. Taking arguments and return values together, you can think of a function as a reusable piece of code that transforms its inputs (arguments) into its outputs (return values).

CREATING FUNCTIONS IN PYTHON

As a reminder, most functions in Python have a name, some optional arguments, some code to run, and an optional return value. Figure 4.1 shows a simple example of a function and points out several of these key components.

The name of the function in Figure 4.1 is "my_function," and that is what we would use to call it. When we call a function, we always use parentheses, as in my_function(). Our example function has one argument, called "my_arg," that we can use to supply some information to the function. When we call a function that receives arguments, we supply the input between the parentheses, like this: my_function("This is a test."). Our example function in Figure 4.1 does a little bit of work—specifically, it calls print to display whatever we have supplied as an argument. Finally, the function ends with a return statement: In this case we return the same thing every time, and that is the value True. Other than serving as an illustration, this is not a particularly useful function, so let's build a real function that can perform a simple but helpful trick: alphabetically sort the column names of a DataFrame. We will need

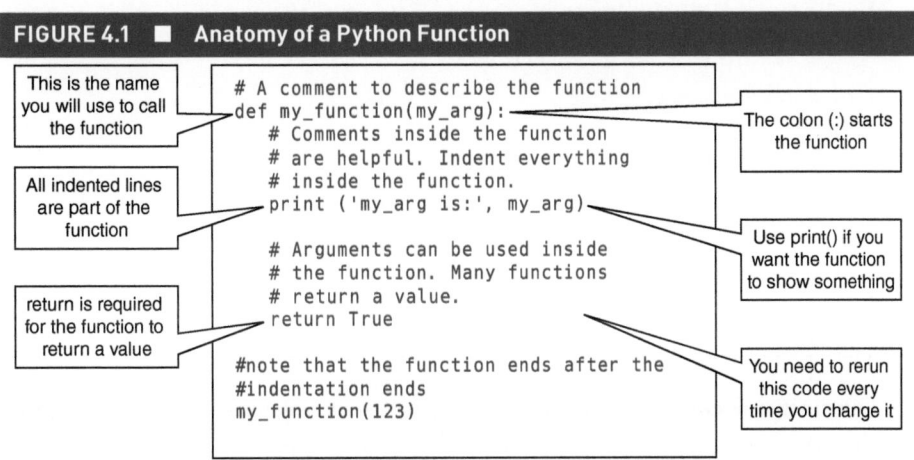

FIGURE 4.1 ■ Anatomy of a Python Function

some input to test the function, so let's first define a simple DataFrame containing information about some companies:

```
data = [['A_corp', 43, 'hotel', 8],
    ['B_corp', 42, 'hotel', 6],
    ['C_corp', 55, 'restaurant', 23],
    ['D_corp', 20, 'restaurant', 9],
    ['E_corp', 89, 'hotel', 20]]

columns_names = ['name', 'sales',
                 'type_of_business', 'profit']

df = pd.DataFrame(data, columns=columns_names)
df

      name    sales   type_of_business    profit
0     A_corp  43      hotel               8
1     B_corp  42      hotel               6
2     C_corp  55      restaurant          23
3     D_corp  20      restaurant          9
4     E_corp  89      hotel               20
```

Now we can start to define our new function. One way that programmers sometimes do their work is to create the simplest possible version of something first, test it, and then add features a little at a time. Here's the beginning of our function, although it does not do anything useful:

```
def col_name_sort(input_df):
    '''
    Here's some documentation!
    '''
    col_names = input_df.columns.values
    return("Success!")

col_name_sort(df)  # Not part of the function

'Success!'
```

Our function receives a pandas DataFrame as input, reads the column names into a new variable, and then returns a string with the word "Success!" Once you run this code, you have tested your first custom function in Python! Note the keyword "def" (define) at the start of the first line and the colon at the end of the first line of the function. These elements are the signal to Python that we want to define a new function. All the subsequent code that is indented becomes part of the function. At this time, we have two lines of code that are indented, so our function doesn't do much.

Our last line of code in this cell is actually outside the function, which we can tell because it is not indented. That last line tests our new col_names_sort() function. When we call the function to test it, we see the string "Success!" that the function returned. Eventually we will want to return something more useful. By the way, when using Jupyter notebooks, it is a good idea for each function definition to have its own cell. In this case, we put the test of the function (the last line) in the same cell for the convenience of formatting for this book. When you create your own function, we suggest that you put it in a cell by itself. One last thing: When professional programmers write a function, they include some documentation that describes what the function does. When you enclose some explanatory text in triple quotes, as is shown in the previous code cell, it is ignored by Python (but is potentially helpful to humans).

Let's now improve our function. One important point to remember when updating functions is that Python will remember whatever is the most recent version of the function you have designed. In practice, we would only define a function once, but for the code files for this book, we have multiple versions of the function as we keep adding new features. As long as you execute the cells in order, everything should work fine.

For this next improvement of our function, let's sort the column names and return the sorted list of column names. We can also take this opportunity to upgrade our documentation:

```
def col_name_sort(input_df):
    """
    A function to alphabetically sort the column names
    of a pandas DataFrame.
    """
    name_list = input_df.columns.values
    name_list = sorted(name_list)
    return(name_list)

col_name_sort(df) # Not part of the function

['name', 'profit', 'sales', 'type_of_business']
```

As before, we receive a DataFrame as the argument. We store the list of the column names in name_list, then sort name_list alphabetically, then return it to the caller. As before, the last line of code is not part of the function, rather it is a test of the function. The output confirms that the names have been sorted correctly.

We're on our way to doing what we wanted to do, but we have not actually changed anything about the input DataFrame. We can add a line of code to accomplish that now. Note that we have also improved our documentation again and changed our return statement to provide the column-sorted DataFrame to the caller:

```
def col_name_sort(input_df):
    """
    A function to alphabetically sort the column names
    of a pandas DataFrame.

    Arguments: input_df - a pandas DataFrame
    Returns: a pandas DataFrame with the columns alphabetized
    """
    name_list = input_df.columns.values
    name_list = sorted(name_list)
    output_df = input_df[name_list] # Here's the new code
    return output_df
```

The new line of code works because the square brackets notation happily accepts a list of the column names in whatever order we want to supply them. This code cell produces no output because we took away the test code that our previous cells held. Note that our documentation also mentions what arguments are expected and what the function returns. In a new code cell, we can test whether our new function does its job properly:

```
col_sorted_df = col_name_sort(df)

col_sorted_df
```

	name	profit	sales	type_of_business
0	A_corp	8	43	hotel
1	B_corp	6	42	hotel
2	C_corp	23	55	restaurant
3	D_corp	9	20	restaurant
4	E_corp	20	89	hotel

This code appears to have worked as expected: When we call col_name_sort() supplying df as the input argument, the function returns a new DataFrame that we assign to the variable col_sorted_df. When we report col_sorted_df it has all of the original columns and data as the original df, but the columns are now organized in alphabetical order.

DEFENSIVE CODING

Python is a flexible language that allows coders to use a variety of types of data as needed in different situations. In fact, Python is known as a "dynamically typed" language, which means that you do not have to tell Python in advance what kind of data each variable is going to be. One side effect of this flexibility, however, is that the rules governing what types of data you can use in a given operation are not applied until the code is already running! For example, if our new function receives as input something that is not a DataFrame, this will create an error that will cause Python to stop running whatever piece of code it got started on. We designed col_name_sort() to accept a DataFrame as input, but we did not check to make sure that it was true. We can use a conditional test to find out whether the function has received the right kind of error and provide an informative error message if it did not:

```
def col_name_sort(input_df):

    # The conditional test
    if isinstance(input_df, pd.DataFrame):
        name_list = input_df.columns.values
        name_list = sorted(name_list)
        output_df = input_df[name_list]
        return output_df
    else:
        print("Error: Expected DataFrame as input.")
        return None
```

In the previous chapter we saw an example of a conditional test when we were filtering a DataFrame, but this is our first use of the if/else conditional. In the second line of code, right after the comment, you can see the keyword "if" is followed by a logical test. This logical test is performed by "isintance()," which asks the question of whether the argument is an instance of a pandas DataFrame. If the test yields the value True, then the indented lines of code on the following lines are executed. If the test yields the value False, we skip over those lines and go directly to the "else" keyword. The indented lines following the else are then run. You can see that the if clause and the else clause provide different return values to the caller: The if clause returns the sorted DataFrame, whereas the else clause returns None. None is a reserved keyword in Python which essentially means, "There's nothing for me to return." Let's test the new version of our function:

```
my_list = [1,2,3]

col_name_sort(my_list)

Error: Expected DataFrame as input.
```

The output shows that our function provided an error message because we supplied a list as input to the function rather than a DataFrame. The error message could be helpful in debugging a problem in our code, but we have overlooked one subtle, but important, point. When we printed this error message and returned None, whatever code we may include that follows this error message will continue to run. The technical term for this is that we have not "interrupted execution." Although that might be OK in some circumstances, consider what might happen if our return value of None made it into some later code: It might cause a bug or error that we did not expect and might have difficulty pinpointing.

The solution to this problem is to use Python's built-in capability for "raising an exception." Take a look at the last line of code in the following cell:

```
def col_name_sort(input_df):
  if isinstance(input_df, pd.DataFrame):
    name_list = input_df.columns.values
    name_list = sorted(name_list)
    output_df = input_df[name_list]
    return output_df
  else:
    raise TypeError("Expected DataFrame as input.")
```

The last line of code raises a "TypeError," which Python defines as a situation where there is a mismatch between the type of data we wanted and the type of data that was provided. In the parentheses we have provided the same helpful error message as before, but in this case if we make it into the else clause, Python will stop the code from running and will provide a bunch of diagnostic information, concluding with our helpful error message. You should try this yourself in the notebook for this chapter by running the code shown here and then going back a few cells to the test code where we supplied a list instead of a DataFrame as the argument to col_name_sort().

Depending upon how mission critical your function is, you can include any number of tests of the assumptions that the function makes. Python has a total of about 60 different kinds of exceptions that you can raise, and if none of those fit the situation, you can create your own custom error type. For now, most of the custom functions we develop will be simple enough that we won't need these advanced features.

Three other points deserve our attention. First, notice that when we created our own function, it was important to do some testing to make sure the code worked the way we expected. Professional data scientists know that when working on anything related to computers, including spreadsheets, macros, scripts, and applications, you must test your code thoroughly before putting it into production. Second, we introduced new functions in this exercise, such as isinstance(). Where did these come from, and how did we know where to find them and how to use them? Python has so many functions, in so many libraries, that it is difficult to memorize them all. In addition, there's almost always more than one way to do something. It can be quite confusing to create a new function if you don't know all of the ingredients. This is where the professional community comes in. Search online, and you will find dozens of instances where people have probably tried to solve similar problems to the one you are solving, and you will also find that they have posted the Python code for their solutions. These code fragments are free to borrow and test. In fact, learning from other people's examples is a great way to expand your horizons and learn new techniques. Make sure that you give credit to the person who posted the example whenever you borrow their code.

Finally, remember that when you write a new function, you are providing a piece of code that will be reused by your future self or whoever might be tasked with maintaining or upgrading your code. We have mentioned before the importance of comments—at least one comment per cell in a Jupyter notebook plus comments on or before any lines of code that are tricky to read. You will also remember that we included inside our function within triple quotes some documentation that describes how the function works. These function comments work with Python's help system:

```
help(col_name_sort)

Help on function col_name_sort in module __main__:

col_name_sort(input_df)
    A function to alphabetically sort the column names
    of a pandas DataFrame.

    Arguments: input_df - a pandas DataFrame
    Returns: a pandas DataFrame with the columns
             alphabetized

    Reports an error and returns None if the argument
    is not a DataFrame
```

Every well-written function in Python and its packages includes documentary comments similar to what we included in col_name_sort(input_df). This means that you and all other Python programmers can run the help() function to find out what a function expects for inputs, what it produces for outputs, and how it transforms the former into the latter. We said it before and will say it again here: Comments and documentation are a gift to your future self (and possibly others as well).

CLASSES AND METHODS

In the earlier part of this chapter, we demonstrated how to define a new function. A function like col_name_sort() stands on its own: Once we have run its code block to define it, we can call it whenever we have a DataFrame for which we want the columns sorted. Python provides all the capabilities needed to create new functions like col_name_sort(), and as a result computer scientists refer to Python as having features of a "functional language." Some years ago, however, a new idea called "object-oriented programming" was introduced. Object-oriented languages include the capabilities needed to create custom "classes"—little bundles of data and code that can both store information and perform operations on that information.

Object orientation has helped programmers develop reusable capabilities customized for particular configurations of data.

You might be surprised to learn that we have already been using Python's object-oriented capabilities: pandas DataFrames are a custom class that provide us with many unique capabilities for creating and manipulating rectangular datasets. In previous chapters when we called pd.DataFrame(), we were "instantiating" (creating) a new member of the pandas DataFrame class. The resulting object not only stored our data in columns and rows, but it also carried along with it the many "attributes" and "methods" that are needed to modify, repair, and transform our data.

Here's a rundown on the new terminology we will need to remember: A class is a template or plan for creating a structure that contains both data and code. When we want to create a new copy of such a structure, we instantiate the class. The resulting object has attributes—persistent pieces of information about the object that we might need to know—as well as methods, which are functions specifically designed to work on that class. An example of an attribute is the shape of a pandas DataFrame. If we write a line of code that says, my_df.shape, Python will report the number of rows and columns of my_df because that information is stored on instances of the DataFrame class. An example of a method is reset_index(). When we write a line of code that says my_df.reset_index(drop=True), we are calling on a specialized function—custom built for the DataFrame class—that knows how to apply a new set of row indexes for our my_df dataset.

Here's an example to illustrate these ideas:

```
class AlphaDF:
    """
    Custom class to maintain a DataFrame where the
    column names are alphabetized.
    """
    def __init__(self, input_df):
        if isinstance(input_df, pd.DataFrame):
            self.name_list = input_df.columns.values
            self.name_list = sorted(self.name_list)
            self.alpha_df = input_df[self.name_list]
            return None
        else:
            raise TypeError("Expected DataFrame as input.")
```

This code creates a new class called AlphaDF. As we did earlier when we created a new function, we include a little documentation within the triple quotes. Every class has an __init__() function (there are two underscores before the name and two after) that defines the persistent variables that the class possesses as well as any startup actions that need to get done when creating a new instance of the class. You should recognize the code that follows, starting with the if statement: It is nearly identical to col_name_sort(). In fact, the only difference is the use of the word "self." Anything with self in front of it refers to a persistent class variable, that is, a data storage area inside the instantiated class.

By convention, every chunk of the class definition, including the __init__() function, receives a reference to self as an input. That's why the definition of __init__() contains self as the first argument: **def __init__(self, input_df)**. Obviously, though, __init__() also receives as input a DataFrame that the code immediately reorganizes with alphabetized columns (like the col_name_sort() function did).

So far, this AlphaDF class does not seem like any improvement over what we could do previously by calling col_name_sort(). But here's the other half of the class definition:

```
def reverse(self):
    """
    A method to reverse the sort of the columns.
    """
    self.name_list = sorted(self.name_list, reverse=True)
    self.alpha_df = self.alpha_df[self.name_list]
    return None
```

This part of the class definition sets up a new method called reverse(), which can reverse the sort order of the column names. The reverse() method needs only two lines of code. The first line reverse-alphabetizes the list of column names—which, by the way, is stored as an attribute of the class. The second line then revises self.alpha_df to include the columns in the new order. Here's a test of the class that illustrates instantiating a new class member and then using the reverse() method:

```
my_new_class = AlphaDF(df)

my_new_class.alpha_df.columns

Index(['name', 'profit', 'sales', 'type_of_business'], dtype='object')
```

```
my_new_class.reverse()

my_new_class.alpha_df.columns

Index(['type_of_business', 'sales', 'profit', 'name'], dtype='object')
```

In the first code cell, we supply the name of a previously defined DataFrame, "df," when we instantiate a new class member. The new member is stored in my_new_class for future use. In the second line of the first code cell, we inspect the list of columns and find that we have correctly sorted them alphabetically.

In the second code cell, we use the reverse() method on the newly instantiated class member. This is like calling a function, but we supply the name of the class member before the name of the method, like this: `my_new_class.reverse()`. In the second line of code, when we inspect the columns, we can see that they are now in reverse alphabetical order.

That might be the shortest introduction to Python classes ever presented, and we have naturally glossed over many of the details. Creating custom classes is not something that all Python programmers need to do, but the one important thing to remember is that when you hear the word "method" being used in reference to some Python code, what you are hearing about is a function that has been specially designed to work with a class. Much of the time, for example when we are using pandas, we may not even be paying attention to the fact that we are using a specialized class. As long as we remember how to ask for attributes and call on methods, we can take advantage of classes that other people have written.

To recap this chapter, by creating our own function, we learned that functions take arguments as their inputs and provide a return value as their output. A return value can be any Python data object, so it could be a single number, or it could be a set of values (a list) or even a more complex data object like a DataFrame. We can write and reuse our own functions, which we will do again later in the book, or we can use other people's functions by importing their libraries. There is a specialized kind of function, called a "method," which can only be called in reference to a member of a class. An example of a class is a pandas DataFrame, and we now have experience creating new DataFrames and using methods such as reset_index() to perform operations on them. Because so many powerful Python libraries, like pandas, already exist, we can accomplish many data science tasks by taking advantage of classes that others have written.

CASE STUDY: CREATING AND USING A CALCULATE NPS FUNCTION

```
Case Key Points:
 - Define a function that accepts a vector of
   likelihood-to-recommend values, and returns the NPS value
 - Test the function to ensure it is working correctly
 - Use the function to explore the survey dataset
```

We can adapt previous code we created to calculate a Net Promoter Score (NPS) to create a new function for this calculation. As we discussed earlier in the book, NPS is a measurement that some businesses use to quickly summarize the attitudes of a set of consumers toward a product or service that the business offers.

To calculate NPS, we need a list of numbers where each number represents a *likelihood to recommend* rating provided in response to the question: "On a scale of 1 to 10, how likely are you to recommend this [product or service]." What follows is the code that creates a function to calculate NPS:

```
def Calc_NPS(ltr, neutral = [7, 8]):

    if isinstance(ltr, pd.Series):
        ltr = ltr.to_list()

    df_tmp = pd.DataFrame(ltr, columns=['ltr'])

    promoters = df_tmp.ltr[df_tmp.ltr > max(neutral)]
    num_promoters = len(promoters) # compute num of promoters

    detractors = df_tmp.ltr[df_tmp.ltr < min(neutral)]
    num_detractors = len(detractors) # compute num detractors

    total = len(ltr) # Total number of ratings
    nps = (float(num_promoters)/total -
           float(num_detractors)/total) * 100

    return nps
```

The version of the code in the notebook for this chapter contains better documentation than what we have shown here. One thing to notice about this function is that it takes two arguments. The second argument, called "neutral," can be used to specify the neutral zone where a respondent is considered neither a promoter nor a detractor. To make the function easier to use, we have supplied a default value in the function specification. In the first line of the function where we mention neutral, we also use an equal sign and provide the default range of 7–8 (this range is consistent with our initial consideration of NPS scores in the previous chapters). Providing these default values means that when the function is called, it is optional to provide values for neutral. If the call to Calc_NPS() does not provide the neutral argument, then 7 and 8 will be used. Other values can be provided to override these defaults.

After running the code to define the calc_NPS() function, we can test the function:

```
ltr = [9,8,3,9,7,8,9,6,7,8,9]

Calc_NPS(ltr)

18.181818181818183
```

Next let's reuse the Calc_NPS function to explore the small survey dataset that we started to explore in the previous chapter. First, as a reminder, we will redo the analysis of the mean values for LTR by type of travel:

```
small_survey = pd.read_csv("https://"
                           "raw.githubusercontent.com/"
                           "jmstanto/IDSpython/main/"
                           "smallSurvey.csv")

small_survey = small_survey.dropna(subset=['LTR'])

small_survey[['LTR', 'Type.of.Travel']].groupby(
                                     ['Type.of.Travel']).mean()

                         LTR
Type.of.Travel
Business travel        8.135314
Mileage tickets        8.051282
Personal Travel        5.461538
```

In this code, we read in the survey from the CSV file, filter out the rows that have missing data for LTR, and then calculate a mean for each type of travel. Based on this analysis, we can see that, collectively, people who do personal travel have lower LTR as compared to business and mileage rewards travelers. Next, let's calculate NPS for all of the respondents. Don't forget that by leaving out the neutral argument, we are accepting the defaults of 7 and 8 for the neutral zone:

```
Calc_NPS(small_survey.LTR)

9.018036072144293
```

Next, let's calculate NPS broken down by group. This code uses a cool trick by supplying our Calc_NPS() function to the aggregate method on the pandas DataFrame. The agg() method will repeatedly call Calc_NPS() for each of the subgroups:

```
group = small_survey.groupby('Type.of.Travel')
group['LTR'].agg(Calc_NPS)

Type.of.Travel
Business travel         40.264026
Mileage tickets         30.769231
Personal Travel        -57.051282
Name: LTR, dtype: float64
```

These results show that NPS is highest among Business ticket travelers, and it is lowest for personal travelers. In fact, the negative NPS for personal travel is quite problematic, indicating that there are more detractors than promoters in each group. This is an important finding that the company needs to hear: Figuring out how to improve the customer experience for personal travelers should be a priority.

CHAPTER CHALLENGES

1. Create a new function that accepts one argument and returns that same argument. In a separate code cell, write a line of code that tests your new function.

2. Create a new function that accepts one argument. The function should test whether the argument is an integer or not using the isinstance() function. If the incoming argument is an integer, the function should return that number. If the incoming argument is not an integer, the function should return None. In addition to your new function, include a cell with at least one line of code that tests the function.

3. Write and test a new function called my_vector_info() that takes as input a list of numbers and prints the key characteristics of the list, namely, the min, max, and mean of the list. In addition to your new function, include a cell with at least one line of code that tests the function. Reminder: You will need to use print() inside of your function so that it can produce output on the console.

4. Enhance your my_vector_info() function to make sure that you test the incoming list to ensure that it is numeric. In addition to your new function, include at least one line of code that tests the function. Hint: you can use the code isinstance(x, (int, float)) to see if the variable x is a number.

5. Create a function that accepts the name of a CSV file as its input. The function should read in that file using read_csv() and then remove rows that have missing data. Make sure to test your function on a CSV file such as the smallSurvey.csv file used in this chapter.

6. Create a function that reads and cleans the census dataset using the repairs that were accomplished in the previous chapter.

7. Modify the Calc_NPS() function developed in this chapter so that the total count of cases used in the NPS calculation is the sum of the number of promoters and the number of detractors (i.e., leaving out the neutrals). Test your code by redoing the calculations by type of travel as shown in this chapter.

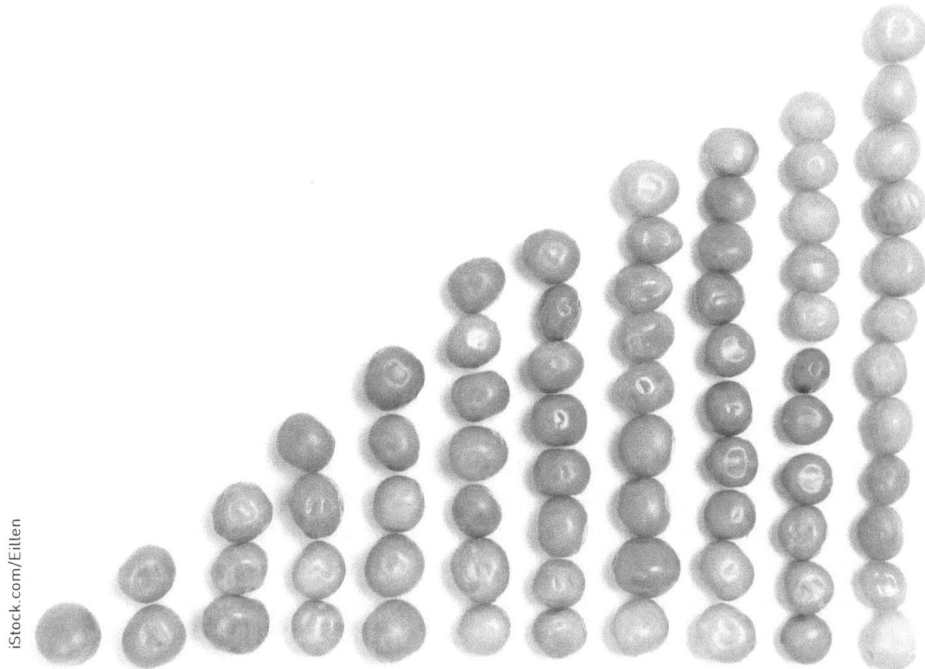

5 BEER, FARMS, PEAS, AND STATISTICS

LEARNING OBJECTIVES

Explain why we sample from a population.

Compare basic descriptive statistics such as mean, median, range, mode, variance, standard deviation.

> Use histograms to explain the concepts of central tendency and measures of dispersion.
>
> Understand generate a normal distribution.
>
> Gain experience calculating statistics using Python code.

HISTORICAL PERSPECTIVE

The end of the 1800s and the early 1900s were a time of astonishing progress in math and science. Given enough time, paper, and pencils, scientists and mathematicians of that age imagined that just about any problem facing humankind—including the limitations of people themselves—could be measured, analyzed, and made more efficient. Four Englishmen who epitomized these idealistic beliefs were Francis Galton, Karl Pearson, William Sealy Gosset, and Ronald Fisher.

First on the scene was Francis Galton, half-cousin to Charles Darwin. Galton was a gentleman of independent means who studied Latin, Greek, medicine, and math and who made a name for himself as an explorer. He applied statistical methods to the study of differences among people and the influence of inheritance, which led to him coining the term "nature versus nurture." Alas, he also introduced the concept of eugenics: the idea that humanity could be improved through selective breeding. Galton studied heredity in peas, rabbits, and people and concluded that certain people should be paid to get married and have children whereas others should be prevented from doing so. These repugnant ideas were later used, most notably by the Nazis during the Second World War, as a justification for killing people based on their ethnic backgrounds. For all his theorizing, Galton was not much of a mathematician, but he had a partner, Karl Pearson, who is often credited with founding the field of mathematical statistics. Pearson refined the math behind correlation and regression and did much more besides to contribute to our modern abilities to manage numbers. Like Galton, Pearson was a proponent of eugenics, but he also is credited with inspiring some of Einstein's thoughts about relativity and was an early advocate of women's rights.

Next was William Sealy Gosset, a wizard of math and chemistry. It was the latter expertise that led the Guinness Brewery in Dublin, Ireland, to hire him. The brewery was on the lookout for ways of making batches of beer more consistent in quality. Gosset developed what we call small sample statistical techniques—ways of generalizing from the results of a relatively few observations. To do this, he had to figure out the role of chance in determining how each batch turned out. The Guinness company frowned on academic publications, so Gosset published under the modest pseudonym, "Student." If you hear someone discussing the "Student's t-Test," that is where the name came from.

Last but not least was Ronald Fisher, a mathematician who also studied biology and genetics. Unlike Galton, Fisher was not a gentleman of independent means. In fact, early in life he and his spouse struggled as subsistence farmers. One of Fisher's later jobs was at an agricultural research farm called Rothhamsted Experimental Station. Here, he analyzed variations in crop yield that led to a widely used statistical technique known as the analysis of variance. Fisher also pioneered the area of experimental design, which includes factors, levels, experimental groups, and control groups that we discuss elsewhere in the book.

Of course, these four are not the only mathematicians and researchers to have made substantial contributions to practical statistics, but they are notable with respect to the applications of statistics to the other sciences, engineering, and agriculture. In some ways, these guys were the grandfathers of data science.

SAMPLING A POPULATION

One of the critical distinctions woven throughout the work of these four is between the sample of data that you have available to analyze and the larger population of cases that might exist. When Gosset ran batches of beer at the brewery, he knew that it was impractical to run every possible batch of beer with every possible variation in recipe and preparation. Gosset knew that he had to run a few batches, describe what he had found, and then infer what might happen in future batches. This is a fundamental aspect of working with all kinds of data: **Whatever data you have, there's always more out there**. There are data that you might have collected by changing the way things are done or the way things are measured. There are future data that hasn't been collected yet and might never be collected. There are data that is inaccessible to us because of the expense or impracticality of collecting them. There are even data that we might have gotten using the exact strategies we did use but that would have come out subtly different due to chance. Whatever data you have, it is only a snapshot, or sample, of what might be out there. This means that we must never, ever 100 percent trust the data we have. There is always uncertainty in data. Much of the usefulness of statistics comes from their capacity to help us understand uncertainty and to guard against putting too much stock in what a sample of data has to say. Remember that although we can always *describe* the sample of data we have, the real trick is to *infer* what the data could mean when generalized to the larger population of data that we don't have. This is the key distinction between descriptive and inferential statistics.

UNDERSTANDING DESCRIPTIVE STATISTICS

We have already encountered descriptive statistics in previous chapters, but here is a more complete list with more-detailed definitions:

> The mean (technically the arithmetic mean), is a measure of central tendency that is calculated by adding together all of the observations and dividing by the number of observations. Many people call this the average.
>
> The median is another measure of central tendency but one that cannot be directly calculated. Instead, you make a sorted list of all of the observations in the sample, then go halfway up that list. Whatever the value of the observation is at the halfway point, that is the median.
>
> The mode is another measure of central tendency. The mode is the value that occurs most frequently in a sample of data. Like the median, the mode cannot be directly calculated. You have to count how many of each number there are and then pick the category that has the most.

The range is a measure of dispersion—how spread out a bunch of numbers in a sample are—calculated by subtracting the lowest value from the highest. Although simplistic, the range provides a quick initial view of dispersion.

The variance, like the range, describes the "spread" of a sample of numbers. Unlike the range, which uses only two numbers to calculate dispersion, the variance is obtained by including all of the numbers in a simple calculation that compares each value to the mean.

The standard deviation is another measure of dispersion and a cousin to the variance. The standard deviation is simply the square root of the variance. As the square root of the variance, the standard deviation is expressed more naturally in the units of the original measurement.

Let's do some simple statistical calculations using the company profit DataFrame that we used last chapter. First, we will put some data in the DataFrame again:

```
data = [['A_corp', 43, 'hotel', 8],
        ['B_corp', 42, 'hotel', 6],
        ['C_corp', 55, 'restaurant', 23],
        ['D_corp', 20, 'restaurant', 9],
        ['E_corp', 89, 'hotel', 20]]

columns_names = ['name', 'sales',
                 'type_of_business', 'profit']

df = pd.DataFrame(data, columns=columns_names)
df
```

	name	sales	type_of_business	profit
0	A_corp	43	hotel	8
1	B_corp	42	hotel	6
2	C_corp	55	restaurant	23
3	D_corp	20	restaurant	9
4	E_corp	89	hotel	20

Here's the code to calculate the mean profit, which you will see is 13.2. The numbers in this column are calibrated in millions of dollars, so keep in mind that when we say that the mean is 13.2, what we are saying is $13,200,000.

```
df['profit'].mean()

13.2
```

Next, we are going to explore the two most popular measures of dispersion, the variance and the standard deviation. For some people, understanding the variance is easier if you can see the calculations in a table.

Table 5.1 shows the row-by-row calculations use to compute variance. On each row, we get the deviations from the mean and then square them (multiplying each one times itself). Note how every squared deviation in the rightmost column is a positive number. The larger the deviation of each original value in the second column from the mean, the larger this squared deviation gets. We add all of the squared deviations and then divide to get a kind of average squared deviation: This is the variance. By the way, it was not a mistake to divide by 4 instead of 5—the reasons for this come from an idea called "degrees of freedom" that affects calculations on samples. When calculating the "average" squared deviation we always divide by n–1, that is, one less than the sample size, to get the correct value for the sample variance.

The result of this calculation, displayed in the table as the value 59.7, is the sample variance, a useful measure of dispersion often used in data analysis. Variance is mathematically valuable, but it is difficult to describe to other people because it is expressed in squared units. For instance, in this example we are looking at an average squared deviation from the mean of 59.7 million dollars. Who measures anything monetary in squared millions of dollars? To address this oddity, statisticians have provided us with the next descriptive statistic, the standard deviation.

In this example, the square root of 59.7 is approximately 7.7 (rounding to one decimal place). That value is the standard deviation, and it is a much more natural statistic to discuss because it is expressed in the same units as the original variable. We can say that the standard deviation of the profit values from our five companies is $7.7 million—that's a phrase that will make sense to more people. Of course, when calculating the standard deviation, we do not usually go through all the rigmarole shown in Table 5.1 . Instead, we call on an existing function or method to do the calculation for us. Let's ask Python to calculate the variance and the standard deviation for us. Pandas has built-in methods for these calculations that can be invoked on any numeric column we wish to name:

TABLE 5.1 ■ Calculation of the Variance

Name	profit	profit-mean	(profit-mean)^2
A_corp	8	8–13.2 = –5.2	–5.2*–5.2 = 27.04
B_corp	6	6–13.2 = –7.2	–7.2*–7.2 = 51.84
C_corp	23	23–13.2 = 9.8	9.8 * 9.8 = 96.04
D_corp	9	9–13.2 = –4.2	–4.2 * –4.2 = 17.64
E_corp	20	20–13.2 = 6.8	6.8*6.8 = 46.24
		Total:	238.8
		Total/4:	59.7

```
print("variance: ", df['profit'].var())
print("std dev: ", df['profit'].std())

variance:     59.7
std dev:   7.726577508832744
```

USING DESCRIPTIVE STATISTICS

That small sample of five companies had so little data that we could have examined all the numbers at a glance without needing any descriptive statistics. More realistic datasets that we often encounter, however, contain so many observations that we need some descriptive tools to provide a useful overview of the data. To illustrate this point, let's bring back our previously examined U.S. population dataset. There's something a little unusual about these state-by-state population data that only becomes evident when we apply descriptive tools to the data. Because we will use this dataset many times, let's first create a function to clean up numbers by reusing some lines of code from Chapter 3. Other than the function definition line, there's nothing new about this code, so let's look more closely at the one thing we did change:

```
def numberize(list_of_number_strings):
    '''
    Removes all commas from text strings of format 123,456
    and coerces to integer. Works on a list of strings.

    Argument: list_of_numbers in a text list
    Returns: a list of integer values
    '''
    output_number_list = \
        [int(element.replace(',', ''))
            for element in list_of_number_strings ]

    return output_number_list
```

In the foregoing code cell, we have defined a custom function called numberize(). The core of this function uses the replace() method to remove all commas from a text string. After that, we call on the int() function to convert the text into a number. We know we need to accomplish this process once for every numeric column in our dataset, so this is the perfect candidate for a chunk of code to place into a function. Not only will this save a lot of future typing, but it also prevents mistakes from creeping into our data processing. Once we have tested numberize(), we can use it repeatedly without any concern that we have copied and pasted code incorrectly. The other nice thing you will notice from reading the documentation is that this function receives a list or a list-like object such as a whole column from a pandas DataFrame. This means that we can process a column of data as shown in this example:

```
one_col_df = pd.DataFrame(["12,345", "67,890"],
                          columns=["text_numbers"])

numberize(one_col_df['text_numbers'])

[12345, 67890]
```

In this testing cell, we create a small pandas DataFrame with one column that contains two rows. We put two pieces of text data into the DataFrame—the text in each case is a five-digit number with a comma separating the thousands. The test showed that the numberize() function correctly turned these numbers into two integers.

Because that worked, we can now create and use read_census() function. The code for this appears in the notebook file for this chapter and matches the set of operations we accomplished in Chapter 3. If you look at the notebook file for this chapter, you will see that the code for read_census() calls the numberize() function four times, once for each numeric column in the census data. The following code runs read_census() and displays a fragment of the resulting DataFrame:

```
US_region_pops = read_census(
    'http://www2.census.gov/programs-surveys'\
    '/popest/tables/2010-2011/state/totals/'\
    'nst-est2011-01.csv')

US_region_pops.iloc[:2,:3]

    state_name    april10census    april10base
0   Alabama       4779736          4779735
1   Alaska        710231           710231
```

We have displayed only the first two rows of the DataFrame and limited the display to the first three columns. It is evident from the display that the numbers have been converted correctly. We also see that the first two states mentioned are Alabama and Alaska, which matches the results we got in Chapter 3.

As a side note, if we moved our custom utility functions, such as read_census() and numberize() into a separate file, we could easily reload and use those functions in future code that we write. With Jupyter notebooks, you can accomplish this by putting these two functions in a separate file called Chapter5Utils.ipynb and then running this line of code in the current file:

```
%run Chapter5Utils.ipynb
```

In a Jupyter notebook, the % character precedes what is known as a "magic." In brief, a magic is a specialized command that can affect the computer environment surrounding your Jupyter notebook. In the example provided here, the %run magic will open the specified notebook and run all of the code cells in that other notebook as if they had been inserted into this notebook. Keep in mind that for this to work in the Google Colab environment, you will also have to upload Chapter5Utils.ipynb to the virtual machine that is running your notebook for this chapter.

Now we're ready to do descriptive statistics with a real dataset. Here are the basic descriptives for the April 2010 population of the U.S. states and territories:

```
print("mean: ", US_region_pops['april10census'].mean())
print("median: ", US_region_pops['april10census'].median())
print("var: ", US_region_pops['april10census'].var() )
print("std: ", US_region_pops['april10census'].std() )


mean:     6009063.980769231
median:       4085220.5
var:      45757914390074.48
std:      6764459.052403299
```

But wait—where is the mode? Pandas does have a statistical mode function, but we have not called it here. This is mainly due to the fact that the mode is useful only as a measure of central tendency in a limited set of situations. Specifically, a column of numeric data needs to have a variety of repeated values in it. For example, the mode of this list [1,2,2,2,3,4] is 2. Our population data, however, has all highly unique integer values. The chances of any two states having exactly the same populations at a particular point in time are practically nil, so there's no point in displaying the mode. If you do need the mode, substitute the work "mode" in place of "mean" in the first line of code. Here's an example:

```
print("mode: ",
      US_region_pops['april10census'].round(-6).mode() )



mode: 0    1000000
Name: april10census, dtype: int64
```

That's an interesting result! Note that we used the round() method, with the argument of −6, which rounds every number to the closest one million. The output shows that if we round to the nearest million, the modal population value—that is, the most commonly occurring population count for a state or territory—is 1 million people. This provides a notable contrast to the mean population value, which was about 6 million people.

USING HISTOGRAMS TO UNDERSTAND A DISTRIBUTION

We now know that the average population of a U.S. region is 6009063.98, with a mode of 1 million and a standard deviation of 6764459.65. Taken in isolation, these values seem confusing. What does it mean when the mean is much larger than the mode? And what are the numbers like when the standard deviation is larger than the mean? These descriptive statistics have all been calculated correctly by pandas, but for most of us, it is difficult to visualize what the underlying set of numbers looks like. What would be much more helpful would be to have a *picture* that shows how a group of numbers are distributed. The word "distribution" is used in several different ways in the world of statistics, but for our purposes we use it to mean a summary or overview of a group of numbers. The picture we need to show the distribution is called a frequency histogram—one of the most common and useful statistical visualizations. Learning how to interpret it can give valuable insights into the nature of a collection of numeric values. Here's how we call for a histogram using functions from a Python add-on package called matplotlib:

```
import matplotlib.pyplot as plt

plt.hist(US_region_pops['april10census'])
plt.savefig('5_1.pdf')
plt.show()
```

Before discussing the actual histogram, let's review the code that we used to create this visualization. First, you can see that we imported a package called matplotlib.pyplot and designated "plt" as its abbreviation. This package is a standard way to generate simple visualizations (in Chapter 7 and Chapter 8 we will explore other packages that generate more complex visualizations). Then, we generated a histogram based on our list of numbers for the April 2010 population. We have saved a copy of the plot to a PDF file that could be included in a report. Finally, we show the output of the plot in the notebook. The resulting visualization appears in this book as Figure 5.1 .

Now that have seen a histogram, it might be helpful to have a little context of what a histogram is. A histogram is a specialized type of bar graph designed to show how often data points occur when they are organized into a set of equal-sized ranges, called "bins." The word "frequencies" here means how many observations fit into a given bin. The X-axis is calibrated to represent the full range of values for the data in question. The width of the bins is automatically adjusted to get a pleasing display that is detailed enough to understand the data but not so spread out as to have many gaps. Experienced data analysts can view a histogram and, at a glance, infer a lot of information about the underlying data. In particular, analysts pay attention to the overall shape of the histogram and the nature of the "tails." A common distribution, known as the "normal distribution," is bell shaped and has symmetric tails showing a gradual drop-off in frequency as you move either left or right of the central values. Asymmetric distributions have a bunch of values on the left side or the right side and a long tail that sticks out in the other direction. Uniform distributions have about the same frequency of data at every point.

The histogram in Figure 5.1 shows an interesting picture of an asymmetric distribution with a long tail pointing to the right. To interpret this graphic properly, you must note the "1e7" that appears in the lower right corner: This is scientific notation signifying that the X-axis is calibrated in tens of millions. This is the first time that we have seen scientific notation in Python. If you haven't used scientific notation before, the way you interpret it is to imagine a value like 0.5 that is multiplied by 10,000,000 (i.e., 10 raised to the power of seven). Numbers like this have so many digits that they are sometimes unwieldy to show on a graph, and that is why scientific notation is used. Keeping this in mind, there are about 26 states or regions with populations of less than 5 million (i.e., 0.5 on the X-axis), another 14 with populations under 10 million, about six states with a population of about 10 million, and then a small number of states with populations greater than 10 million. There's also one outlier state with around 35 million inhabitants.

Let's review how we glean this kind of information from the graph. First, look along the Y-axis (the vertical axis on the left) for an indication of how often the data occur in a given category. The tallest bar is to the right of this, and it is just past the 25 mark. To know what this tall bar represents, look along the X-axis (the horizontal axis at the bottom). The first bar ends just before 0.5, which translates to somewhat less than 5 million; let's call it roughly 4 million. So, each new bar (or an empty space where a bar would go) goes up by about 4 million in population. With these points in mind, it should now be easy to see that there are 26 states with populations up to 4 million.

If you think about presidential elections, or the locations of schools and businesses, or how a single U.S. state might compare with other countries in the world, it is interesting to know that there are a few big states in the U.S. and then many much smaller states and regions. Once you have some practice reading histograms, all of the knowledge becomes available at a glance. In contrast, there is something unsatisfying about this diagram. The X-axis is calibrated to have tick marks every 5 million, but the bars don't exactly match this amount. We have 10 bars here to span a range of zero to a maximum slightly higher than 35 million. Why did Python choose to display 10 bins, and is there any way we can clean this up?

The answer is that the hist() function has an algorithm, or recipe, for deciding on the number of categories or bins to use by default. The number of observations, the spread of the data, and the amount of empty space there would be are all considered. Fortunately, it is possible and easy to tell matplotlib to use a specific span for each bin, like this:

```
five_mil_bins = range(0, 45000000, 5000000)
plt.hist(US_region_pops['april10census'], bins=five_mil_bins)
plt.savefig('5_2.pdf')
plt.show()
```

| FIGURE 5.1 ■ Frequency Histogram of April 10, 2010, Census Population Values |

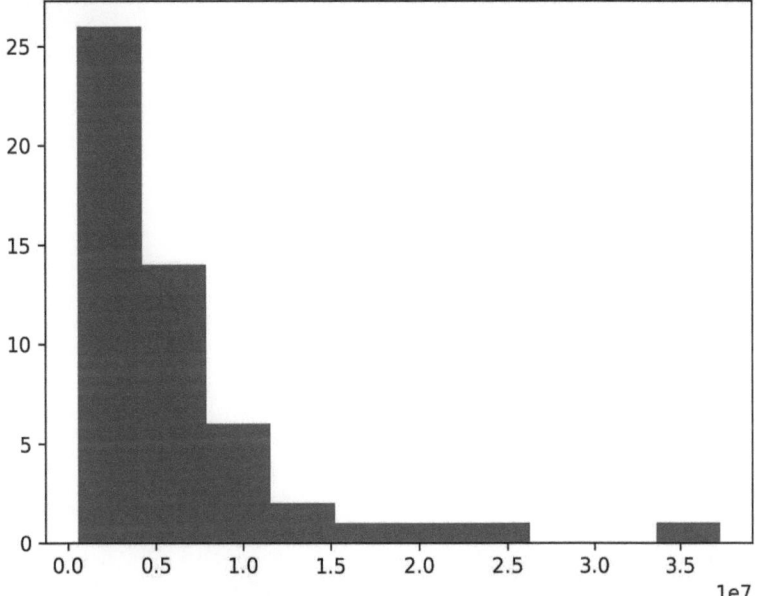

This code uses the range function to create a sequence of numbers. The first two values designate the start of the sequence and the end of the sequence (expressed as the next highest value that will not appear in the sequence). The third value is the step size, which we have set to be exactly 5 million. The result appears in Figure 5.2 .

The histogram in Figure 5.2 is easier to interpret. The first bar shows that there are exactly 30 states or territories with populations of 5 million or less. In the right-hand tail we can also now more clearly see that there are two outlier states: one with a population between 25 and 30 million and one with a population between 35 and 40 million. The overall pattern that you see in both Figures 5.1 and 5.2 starts off tall on the left and swoops downward quickly as it moves to the right. You might call this a reverse-J distribution because it looks a little like the shape a J makes although flipped around. Technically speaking, a statistician might call this a Pareto distribution, named after the civil engineer and economist Vilfredo Pareto, who was interested in the distribution of wealth among people in a society.

We can make some intelligent guesses on why the distribution of state populations looks the way it does. First, it does not make sense to have a state with zero people in it, and there's no such thing as a negative number for a population value. A state has to have at least a few people in it, and if you look back through U.S. history, every state began as a colony or a territory that had a small number people in it. Contrastingly, what does it take for a state or region to grow to become large in population? You need many places to live, first, and then a good reason for people to live there or move there. In other words, there are many limits to growth: Rhode Island is too small in land mass to be likely to have millions of people in it, and Alaska, although it has massive amounts of land, is too cold in the winter for most people to want to

FIGURE 5.2 ■ Frequency Histogram of April 10 Population With Bin Size of 5 Million

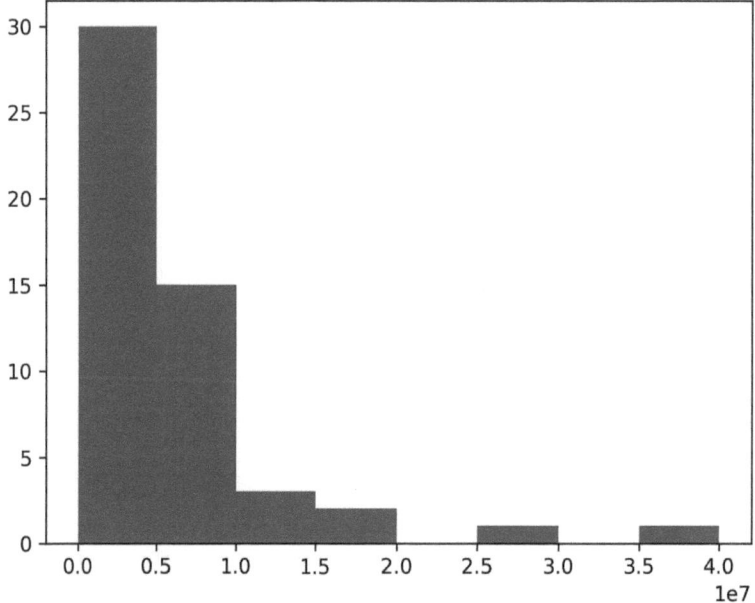

move there. Thus, although all states and territories started small and grew, it is difficult for any area's population to grow huge unless there is a lot of land and powerful reasons for people to live there. As a result, we have a distribution where most of the cases are clustered near the bottom of the scale and only a few are near the top. As you go higher, there are fewer states that have grown really large, and by the time you are at the top end of the scale, shy of 40 million people, there's only one state that has managed to get that big. By the way, do you know or can you guess what that humongous state is? Next we show a line of code with a conditional expression that reveals the largest state:

```
US_region_pops['state_name'] \
     [max(US_region_pops.april10census)==
          US_region_pops.april10census]


4      California
Name: state_name, dtype: object
```

California is the fourth state in our list alphabetically and is the state with the highest population. We used a conditional expression that looks for a match between the maximum value in the april10census column and all of the values in that column. That conditional creates a map of True and False values, and the rest of the expression displays the value for state_name in the one case where the map of True/False contains True.

At this point we finally have the answers to the questions we originally posed about the mean, mode, and standard deviation. Population values for the U.S. states and territories display a right-skewed distribution, possibly a Pareto distribution, with a large cluster of relatively small states and a quite small number of really large states. The mode of such a distribution, assuming we round our values appropriately, will be represented by the tallest bar of a histogram, often appearing at or close to the far left of the frequency distribution. In a distribution with such a long tail, the mean will be dragged outward in the direction of the tail and will therefore represent a compromise between the cluster of small states and the outsized influence of those few large states. The large states will also have a substantial effect on the standard deviation: The squared deviations for those state population values far from the mean will vastly inflate the calculation of the sum of squared deviations, thereby making the standard deviation quite large relative to what would be expected in a symmetric distribution.

NORMAL DISTRIBUTIONS

There are many other distribution shapes besides the Pareto shape shown in Figures 5.1 and 5.2. A common one that almost everyone has heard of is sometimes called the bell curve because it is shaped like a bell. The technical name for the bell curve is the "normal distribution." The term "normal" was first introduced by Carl Friedrich Gauss (1777–1855), who supposedly named it in a belief that it was the most typical distribution of data that one might find in natural phenomena. The histogram in Figure 5.3 depicts the typical bell shape of the normal distribution.

The data in this histogram were generated by the random library, which offers a variety of methods for generating random numbers. In this case, we have called the normalvariate() function, which generates one random number. When we call this function many times, the resulting set of numbers will begin to fit a normal distribution. The numbers will match a normal distribution more closely if you generate a lot of data but less closely if you generate only a small amount. The command used to generate this histogram was as follows:

```
import random
values = \
    [random.normalvariate(6009063.98, 6764459.65)
        for i in range(5000) ]

plt.hist(values, bins=20)
plt.savefig('5_3.pdf')
plt.show()
```

If you look closely at the call to normalvariate(), you will see that we used the mean and standard deviation that were calculated from the U.S. census data. The data shown in Figure 5.3 show approximately what the distribution of population values might look like if, instead of being a Pareto distribution, they were normally distributed. Of course there are negative values, because the distribution is symmetric around the mean, but negative population values are nonsensical.

FIGURE 5.3 ■ A Random Normal Distribution

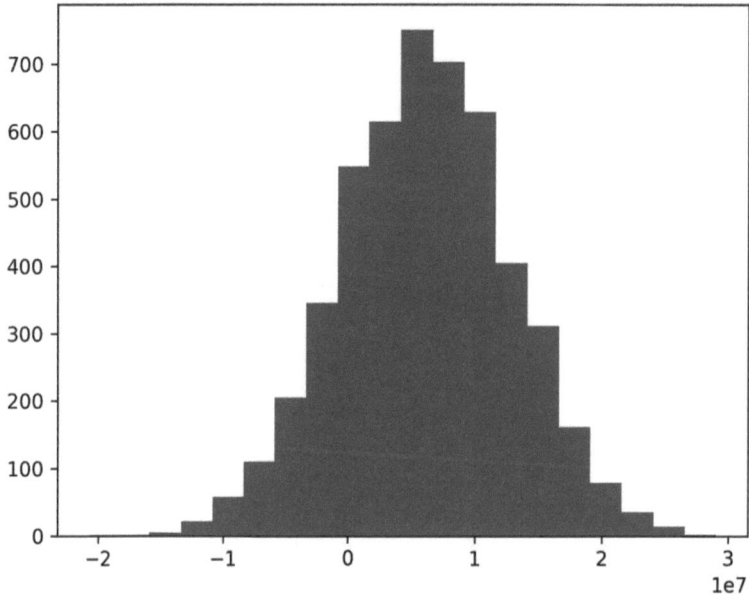

Distributions are used extensively through applied statistics as a tool for making comparisons. We already know from our real state population data that there is only one actual state with a population in excess of 30 million (as you saw from our code, it is California). So, if all of a sudden, someone mentioned to you that they lived in a state, *other than California*, that had 30 million people, you would automatically think to yourself, "Wow, that's unusual, and I'm not sure I believe it." And the reason that you found it hard to believe was that you had a distribution to compare it to. Not only did that distribution have a characteristic shape, but it also had a center point, which was the mean, and a spread, which in this case was the standard deviation. Armed with those three pieces of information—the type or shape of distribution, an anchoring point such as the mean, and a measure of dispersion—you have a powerful tool for making comparisons.

CASE STUDY: EXPLORING LTR DISTRIBUTIONS

```
Case Key Points:
- Use histograms to visually explore the
  distribution of LTR, grouped by Type of Travel
```

Rather than only calculating NPS as we did earlier, let's dig a little deeper into the LTR values. In particular, let's look at the distribution of LTR values across two key groups – business and personal travel.

```
smallSurvey = pd.read_csv("smallSurvey.csv")

smallSurvey = smallSurvey.dropna(subset=['LTR'])

bus_travel = \
    smallSurvey.LTR[smallSurvey['Type.of.Travel']
                    == 'Business travel']

per_travel = \
    smallSurvey.LTR[smallSurvey['Type.of.Travel']
                    == 'Personal Travel']

len(bus_travel), len(per_travel)

(303, 156)
```

In this code cell, we use similar code as before to read in the CSV file containing the small airline dataset. We then remove any missing data from the resulting pandas DataFrame based on any rows that do not contain a value for LTR. In the next two lines of code, we use conditionals to select LTR for subsets of rows, first those rows designated for business travel and, second, those rows designated for personal travel. We will now generate a frequency histogram—actually two frequency histograms overlaid on one another (Figure 5.4).

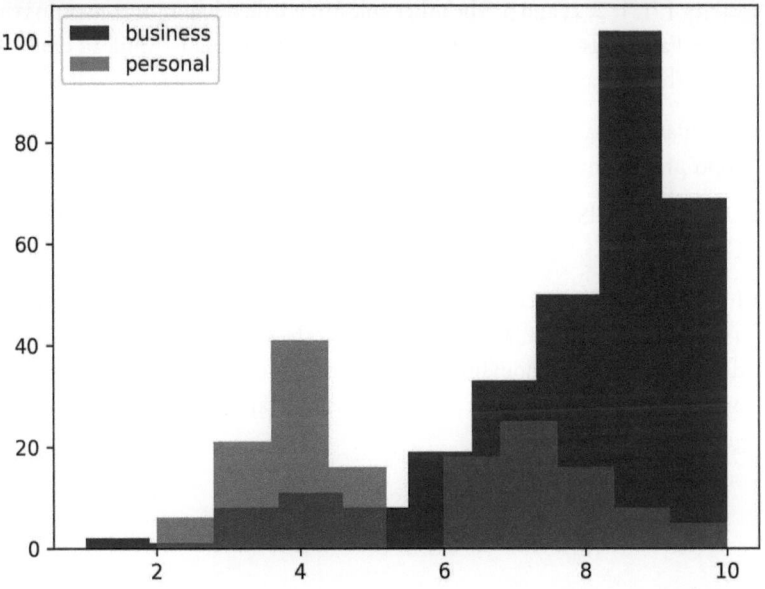

FIGURE 5.4 ■ Two Overlaid Histograms of Likelihood to Recommend (LTR) Scores

We used several features of matplotlib to get this done. First, for each histogram, we specified an alpha value to control the transparency of the plotted bars. Transparency ranges from 0 to 1. A 0 value for alpha makes the color fully transparent (you do not see it), and a value of 1 makes the color opaque (not transparent). The second feature we used was to generate both histograms prior running the code to show the plot. We specified the color of the first histogram as black—the second one defaulted to grey. Finally, we created a legend in the upper right part of the histogram visualization. Here is the code:

```
plt.hist(bus_travel, alpha=0.75, label='business', color="black")
plt.hist(per_travel, alpha=0.75, label='personal')
plt.legend(loc='upper left')
plt.savefig('5_4.pdf')
plt.show()
```

Using an overlay is a valuable way to compare two distributions. In exploring the two histograms, we can see that there are more business travelers who were promoters (specifically 9s), and more personal travelers who were detractors (specifically 3s, 4s, and 5s). The fact that there are many more personal travelers who are detractors, suggests they were unhappy about some aspects of their flights. These factors need to be identified and fixed if the airline wants to improve NPS results.

CHAPTER CHALLENGE

1. Create a small DataFrame that has one column called my_rating that contains the values 1, 1, 1, 2, 2, and 5. The other column is called happy and contains the values True, False, False, True, True, False. Before you calculate the mean of my_rating, using the pandas mean() method, can you guess what the mean will be?

2. Calculate the variance and the standard deviation of the my_rating column using the appropriate pandas methods.

3. Using the smallSurvey dataset, create overlaid histograms of LTR grouped by Gender.

4. In this chapter, we used normalvariate() to generate random numbers that closely fit a normal distribution. We also learned that the state population data was a Pareto distribution. The random library has a function called paretovariate(), which takes one argument called "alpha." Supply a positive value of alpha, such as 1, 2, or 3, and generate 5,000 numbers.

5. Create a histogram of the random numbers from the previous exercise, and write a comment describing its shape.

6. Make a copy of the read_census() function and the numberize() function from this chapter. Test these functions on a different CSV file from the census. You can find a listing of other similar files at this URL: https://www2.census.gov/programs-surveys/popest/tables/2010-2011/state/totals/ If either of your functions causes an error or warning, modify the code to take care of the problem.

iStock.com/natalie-claude

6 SAMPLE IN A JAR

LEARNING OBJECTIVES

Create and interpret sampling distributions.

Use Python to sample repetitively.

> Understand the effects of randomness when using samples.
>
> Demonstrate the law of large numbers and the central limit theorem.
>
> Gain experience using Python statistical functions.

Imagine a jar full of gumballs of two colors: red and blue. The jar was filled from a source that provided 100 red gumballs and 100 blue gumballs, but when they were all poured into the jar, they got mixed together. If you drew eight gumballs from the jar at random, what colors would you get? If things worked out perfectly, which they rarely do, you would get four red and four blue, the same ratio of red and blue that is in the jar as a whole. Of course, it rarely works out this way, does it? Instead of getting four red and four blue, you might get three red and five blue or any other mix you can think of. In fact, it would be possible, although perhaps not likely, to get eight red gumballs. The basic situation, though, is that we don't know what mix of red and blue we will get with one draw of eight gumballs. That's the force of randomness affecting our sample of eight gumballs in unpredictable ways.

Here's an interesting idea, though, that is no help at all in predicting what will happen in any one sample but is great at showing what will occur *over the long run*. Pull eight gumballs from the jar, count the number of red ones, *and then throw them back in* (this is an important procedure known as sampling with replacement). Mix up the jar again, then draw eight more gumballs, and count the number of red. Repeat this process many times. Table 6.1 contains an example of what you might get for the first four draws.

Notice that the left-hand column is a row index showing each time that we drew a new sample. The right-hand column is the interesting one because it is the count of the number of red gumballs in each particular sample draw. In this example, results are all over the place. In sample draw 4 we had only two red gumballs, but in sample draw 3 we had six. But the most interesting part of this example is that if you *average* the number of red gumballs over all of the draws, it comes out to *four red gumballs* per draw, which is what we would expect in a jar that is half red and half blue. Now this is a contrived example, and we won't always get such a perfect result so quickly, but if you tallied 4,000 draws instead of four, you would get an average of almost exactly four red gumballs per draw.

This process of repeatedly drawing a subset from a population is called sampling. Note that we are using the word "population" here in its statistical sense to refer to the totality of

TABLE 6.1 ■ Four Sample Draws From a Gumball Jar

Draw	# Red
1	5
2	3
3	6
4	2

units from which a sample can be drawn. The population in this example is the whole jar of gumballs. If we do a lot of sampling and record the results each time (i.e., the number of red gumballs), the end result is called a sampling distribution.

Next, let's have Python help us draw samples from our U.S. state dataset. As a reminder, you can take a quick look at the start of Chapter 5 to see how we read in the U.S. population and created a DataFrame that contains the population of each state. Also, it is a coincidence that our dataset includes a variable for the number of people in each state and that this value is referred to as "population." The term "population," when used in the statistical sense, does not have to be about people; it signifies a complete set of things (e.g., people, cars, bank loans) from which samples can be drawn.

SAMPLING IN PYTHON

Python has a function called choices(), which is in the random library, that will draw a random sample of a specified size from any dataset with one function call. To explore this function, we first load pandas, random, and statistics libraries. In addition, we will need to do a bit of math (square root), so we include the math library. Finally, we load matplotlib, which will allow us to do some basic visualizations.

The function definitions for numberize() and read_census() are included in the notebook for this chapter, but they are in a hidden code cell. Make sure that the code in that cell has run before running the following:

```
US_region_pops = read_census(
    'http://www2.census.gov/programs-surveys'\
    '/popest/tables/2010-2011/state/totals/'\
    'nst-est2011-01.csv')

US_region_pops.iloc[:2,:3]

     state_name    april10census    april10base
0    Alabama       4779736          4779735
1    Alaska        710231           710231
```

This confirms the first two rows of data—same as we have seen before. Once we have the dataset loaded and repaired, we can sample our census data using the choices() function. Choice is a function inside random library, so don't forget to import random library function:

```
import random
random.seed(1)

random.choices(US_region_pops['april10census'], k=8)

[3574097, 2763885, 1052567, 12830632, 5988927, 5303925, 9535483, 814180]
```

We've introduced a new function here, called "random.seed()." This function controls the random number generation process in Python so that you can obtain the same results as we did. Functions like choices() that use random numbers can be hard to debug without a way to make the results more predictable. When we use random.seed(1) the sequence of random numbers provided will be the same each time. The argument 1 is arbitrary: As long as everyone who is comparing results uses the same seed value, you will get reproducible results.

In the next line of code, the first argument for the choices() function is the source of the data to sample from: We have provided the april10census column from the DataFrame. For the second argument, we have used a named argument: The k=8 controls the number of observations we want Python to randomly sample from the data we provided. The name "k" can be a little confusing: As the sample size, this is the same value that statisticians refer to as "n." Note that the choices() function is used for random sampling with replacement. As noted, this is the procedure recommended by statisticians for mathematical reasons. Generally speaking, for larger samples, the choice of sampling with or without replacement doesn't matter. For small samples, for example, fewer than 30 observations, we should always use sampling with replacement.

When we're working with variables like april10census, instead of counting gumballs of a specific color, we're interested in computing the mean for each sample we draw. We can add that calculation to our code:

```
statistics.mean(\
   random.choices(\
     US_region_pops['april10census'],\
     k=8))

10995086
```

The output no longer shows the eight values that Python sampled from the complete list. Instead, the code used those eight values to calculate the mean. From Chapter 5, you may remember that the actual mean of our observations is 6,009,064. That's notable: The mean that we computed from this one sample of eight states is not even close to the true mean value of our complete set of observations. But there's no need for concern about this result. We know that when we draw a sample, whether it is gumballs or state census values, we will rarely hit the true population mean right on the head, unless we draw a very large sample.

A REPETITIOUS SAMPLING ADVENTURE

The next phase is not something that we would typically do in a practical analysis. Instead, we want to explore some properties of the sampling process as a stepping stone toward understanding how to make decisions with statistics. We will do this by intentionally replicating the sampling process. When we are trying to understand what happens in the long run, we're not interested in any one sample or sample mean but rather in what happens when we replicate the process of sampling many times. Keep in mind that we are working in this artificial situation where we have ready access to the complete census data for all 52 states and territories (what statisticians would call the population), but we are pretending that we don't by drawing many little samples and seeing how far away they are from the real mean. We will ask Python to repeat the sampling process for us, not once, not four times, but 400 or 4,000 times. To start, let's try four replications:

```
[ statistics.mean(
    random.choices(
        US_region_pops['april10census'], k=8) )
            for i in range(4) ]


[2793834, 3623274, 8325464, 3798273]
```

As you can see, we took the exact code as before, which was a nested function to calculate the mean of a random sample of states. This time, we put that code inside the brackets and added a "for loop" to make a list comprehension, so we could run it over and over again. We ran it only four times so that we would not have a big screen full of numbers. From here, though, it is easy to repeat the process 400 times:

```
list_of_means = [ statistics.mean(
    random.choices(US_region_pops['april10census'], k=8) )
        for i in range(400) ]

overall_mean = statistics.mean(list_of_means)

overall_mean

5911317
```

This code block accomplishes the following: (a) draws 400 random samples of size n=8 from our full dataset of states, computes a mean for each one, and assigns the resulting list to the variable list_of_means; (b) when finished creating and saving the list of 400 sample means, the final line of code calculates the overall mean of that list of means. The resulting value should be close to the real april10census mean of 6,009,064. We're off by about 97747, which is roughly an error of 1.6 percent (more precisely, 97747/6009064 = 1.6%). Let's push this repetitive sampling process farther and see if we can get closer to the true mean for all of our data:

```
list_of_means = [ statistics.mean(
    random.choices(US_region_pops['april10census'], k=8) )
        for i in range(4000) ]
```

```
overall_mean = statistics.mean(list_of_means)

overall_mean
```
6030583

Now we are even closer to the true value. If we went up to 40,000 replications, we could get even closer. By the way, don't worry if you get slightly different numbers. From time to time Python changes the random number generation code, so you could easily get slightly different results. The important principle holds, however: The more samples you draw and average, the closer you will get to the true value from the complete dataset.

Before ending our repetitious sampling adventure, let's generate a visual depiction of our process. Instead of summarizing our whole sampling distribution into a single average, we will examine the entire distribution of means using a histogram. Let's also crank up the number of replications to get an even better result than before. The histogram In Figure 6.1 displays a list of 40,000 means as frequencies. Take a close look so that you can get more practice reading frequency histograms.

The histogram shows a typical configuration that is nearly bell shaped, although it has a bit of skewness off to the right. The skewness is caused by having either or both the two very

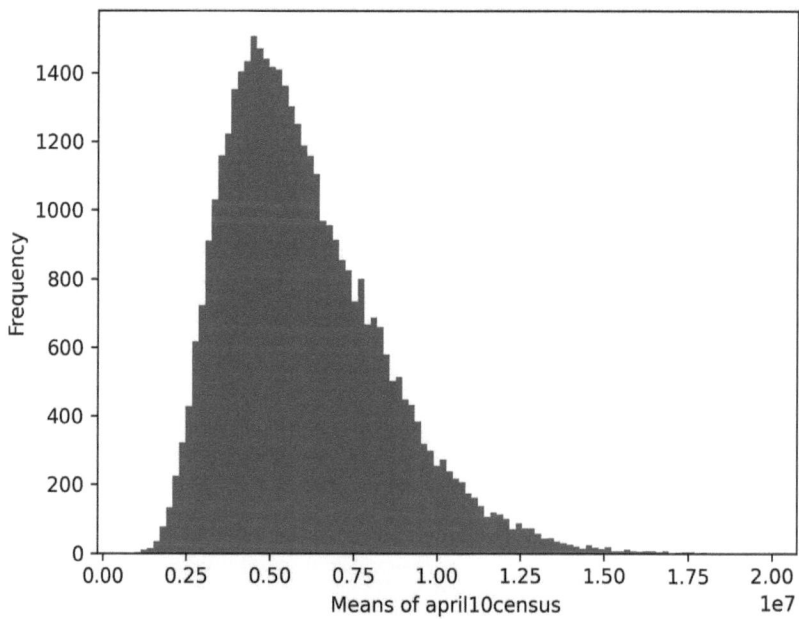

large states (California and Texas) in some of the samples. The tallest, and therefore most-frequent, range of values is near the true mean of 6,009,064.

The implications of this histogram bear careful consideration. First, if you sample repeatedly from a dataset, such as the April 2012 census data, you will get a somewhat different sample mean each time. If you repeat that process many times, and look at the distribution of those sampling means, it will be bell shaped. Furthermore, the center of that bell-shaped distribution will be almost identical to the true population mean, in this case the value 6,009,064 of the April 2010 data.

Now that you understand the histogram, let's take a closer look at the code we used to create it. First, we generate the sampling distribution in the same way as before:

```
list_of_means = [statistics.mean(\
    random.choices(\
    US_region_pops['april10census'],\
    k=8) ) for i in range(40000) ]
```

Then we generate a histogram using matplotib, similar to what we have done previously:

```
plt.hist(list_of_means, bins=100)
plt.xlabel("Means of april10census")
plt.ylabel("Frequency")
plt.savefig('6_1.pdf')
plt.show()
```

As with everything in Python, there are many options for creating a histogram. In Chapter 5 we controlled the bin width; here we are setting the number of bins to 100.

LAW OF LARGE NUMBERS AND THE CENTRAL LIMIT THEOREM

This is a great moment to take a deep breath. We've covered a couple hundred years of statistical thinking in only a few pages. In fact, there are two big ideas here: the law of large numbers and the central limit theorem, which we have partially demonstrated. These two ideas took mathematicians including Gerolamo Cardano (1501–1576) and Jacob Bernoulli (1654–1705) several centuries to figure out.

If you look these ideas up, you might find many complex mathematical details, but for our purposes, there are two important takeaway messages. First, if you run a statistical process a large number of times, it will tend to converge on a stable result (the law of large numbers). For us, we actually already knew what the average population was in our whole dataset. These observations were our population, and we wanted to know how many smaller subsets, or samples, of size n=8 we would have to draw before we could get a good approximation of that true value. We learned that drawing one sample provided a poor result. Drawing 400 samples gave us a mean that was off by 1.6 percent. Drawing 4,000 samples and then 40,000 gave us means that were off by even less than that. If we had kept going to 400,000 repetitions of our sampling process, we would have come extremely close to the actual average of 6,009,064.

Second, we find that the distribution of sampling means starts to create a bell-shaped or normal distribution, and the center of that distribution—the mean of all of those sample means—gets close to the actual population mean. That's the heart of the central limit theorem. The mean of the sample means gets closer faster for larger samples; in contrast, for smaller samples you have to draw lots and lots of them to get close. For fun, let's illustrate this with a sample size that is larger than 16. Here's a run that repeats only 100 times, but each time draws a sample of n=52 (equal in size to the population):

```
statistics.mean(
    [ statistics.mean(
        random.choices(US_region_pops['april10census'],
            k=52) ) for i in range(100) ])

5943435
```

Remember, your result may be slightly different due to randomness. Now, with only 100 samples of n=8, we're off from the true value of the population mean by only a little bit. You might be scratching your head now, saying, "Wait a minute. Isn't a sample of 52 the same thing as the whole list of 52 observations?" This does seem confusing, but it goes back to the question of sampling with replacement that we examined earlier. Sampling with replacement means that as you draw out one value to include in your random sample, you immediately chuck that observation back into the list so that, potentially, any one of them could get drawn again immediately or later. In fact, we could draw samples larger than our population size with no trouble:

```
statistics.mean(
    [ statistics.mean( \
        random.choices(US_region_pops['april10census'], \
            k=5200) ) for i in range(100) ])

6007019
```

That command runs 100 replications using samples of size n=5,200. Look how close the mean of the sampling distribution is to the population mean now. Remember that this result will change a little every time you run the procedure because different random samples are being drawn for each run. But the rule of thumb is that the bigger your sample size, what statisticians call n, the closer your estimate will be to the true value. Likewise, the more trials you run, the closer your population estimate will be.

MAKING DECISIONS WITH A SAMPLING DISTRIBUTION

So, if you've had a chance to catch your breath, let's move on to making use of the sampling distribution. First, let's save one distribution of sample means so that we have a fixed set of numbers to work with. We reset the randomization seed here to start with a clean slate:

```
random.seed(1)

sample_means = [ statistics.mean(
    random.choices(US_region_pops['april10census'], k=8) )
        for i in range(10000) ]

sample_means[:5]

[5232962, 10995086, 2793834, 3623274, 8325464]
```

We're saving a distribution of 10,000 sample means, each of size n=8, to a new vector called sample_means. We'll immediately convert this to a pandas Series because that will give access to a variety of helpful functions. A Series is like a single column extracted from a DataFrame, so we can use helpful methods like describe() on it:

```
mean_series = pd.Series(sample_means)

mean_series.describe()

count    1.000000e+04
mean     5.994790e+06
std      2.342784e+06
min      7.788100e+05
25%      4.231574e+06
50%      5.580999e+06
75%      7.369493e+06
max      1.876124e+07
dtype: float64
```

The output of describe() shows that there are 10,000 observations with a mean value of 5,994,790, which is close to the true April 2010 mean of 6,009,064. We also get readouts of the standard deviation, minimum value, quartiles, and maximum value. The minimum sample mean in the list was 778,810. Think about that for a moment. How could a sample have a mean that small when we know that the true mean is much higher? Rhode Island must have been drawn several times in that sample! The answer comes from the randomness involved in sampling. If you run a process 10,000 times, you are definitely going to end up with a few weird examples. It's almost like buying a lottery ticket. The vast majority of tickets are not a winner. Once in a great while, though, there is an unusual ticket—a winner. Sampling is the same: The extreme events are unusual, but they do happen if you run the process enough times. The same goes for the maximum: At nearly 19 million, the maximum sample mean is much higher than the true mean.

At 5,580,999 the median (the 50% quartile) is close to the mean but not exactly the same because we still have a little bit of rightward skew. The median is useful because it divides the sample exactly in half: 50 percent, or exactly 5,000 of the sample means, are larger than 5,580,999, and the other 50 percent are lower. So, if we were to draw one more sample from the population, it would have a 50–50 chance of being above the median.

Quartiles help us cut things up even more finely. As a reminder, the first quartile is the value that divides the first quarter of the cases from the other three quarters. The third quartile divides up the bottom 75 percent from the top 25 percent. We can see that 25 percent of the sample means are higher than 7,369,493. That means that if we drew a new sample from the population, there is only a 25 percent chance that it will be larger than that. Likewise, in the other direction, the first quartile tells us that there is only a 25 percent chance that a new sample would be less than 4,231,574. If you need a refresher on the median and quartiles, take a look at Chapter 2.

Although the quartiles are useful, statisticians advise us to divide our sample into three regions: A central area containing 95 percent of the sample means, a lower tail containing 2.5 percent of the sample means, and an upper tail that also contains 2.5 percent of the sample means. The reasons for this have a lot to do with habit and tradition, but the basic goal is to divide the "usual" (the central region) from the "unusual" (the upper and lower tails). Pandas can compute these thresholds for us:

```
mean_series.quantile([0.025, 0.975])

0.025    2.577981e+06
0.975    1.143317e+07
dtype: float64
```

This result shows that if we drew a new sample, there is only a 2.5 percent chance that the mean would be lower than 2,577,981. Likewise, there is only a 2.5 percent chance that the new sample mean would be higher than 11,433,171 (because 97.5% of the means in the sampling distribution are lower than that value). Considered from another angle, about 95 percent of all of the samples we drew fell in the range of 2,577,981–11,433,171. Only about 5 percent of all sample means in our sampling distribution are more extreme (i.e., outside of that range). These quantile thresholds therefore provide a basis for decision-making: Any sample means that we observe lower than 2,577,981 or higher than 11,433,171 would be considered highly unusual—possibly indicating that they were drawn from another source, rather than from our list of U.S. states and territories.

EVALUATING A NEW SAMPLE WITH THRESHOLDS

Now let's put this knowledge to work. Here is a sample showing the number of people in eight different areas, where each of these areas is some kind of a unit associated with a country (e.g., a district, county, state, province, or region):

```
1827937  1169936  1184996  2665018    2662480    1325578    2767761    2585518
```

We can enter these into Python and calculate the sample mean:

```
mystery_sample = [1827937, 1169936, 1184996, 2665018,
                  2662480, 1325578, 2767761, 2585518]

statistics.mean(mystery_sample)

2023653
```

The mean of our mystery sample is 2,023,653. Here's the question: Is this a sample of U.S. regions, or is it something else? Looking at it all by itself, it would be hard to tell. Some of these observations have larger values than some U.S. states, and some do not. Thanks to the work we've done earlier in this chapter, however, we have an excellent basis for comparison. We have the sampling distribution of means, and if the new mean is way out in the extreme areas of the sample distribution, say, below the 2.5 percent mark or above the 97.5 percent mark, then it seems much less likely that our mystery sample is a sample of U.S. states and territories. In this case, we can see quite clearly that 2,023,653 is on the extreme low end of the sampling distribution. Recall that when we ran the quantile() command, we found that only 2.5% of the sample means in the distribution were smaller than 2,577,981. The mystery sample seems too small to be a likely member of our sampling distribution.

And this is in fact correct: Our mystery sample contains the number of people in a sample of states from Mexico (using 2010 population values). The important thing to take away is that the characteristics of this group of data points, notably, the mean of this sample, was sufficiently different from a known distribution of means that we could make an inference that the sample *was not drawn from the original population of data*. This method of reasoning is the basis for one of the standard methods of statistical inference. You construct a comparison distribution based on some statistical values, you mark off a zone of extreme values, and you compare any new sample of data you get to the comparison distribution to see if it falls in the extreme zone. If it does, you tentatively conclude that the new sample was obtained from some source other than what you used to create the comparison distribution. Statisticians use particular distributions—for example the "t" distribution—when constructing their tests, so our informal demonstration here does not show the whole picture. As long as you have understood the essential strategy we have shown here, however, then you are well on your way toward grasping the logic of statistical inference.

Later in the book we will come back to specific statistical procedures that use the reasoning described here. For now, we need to take note of three additional pieces of information. First, we looked at the mean of the sampling distribution, but we did not pay any attention to the standard deviation, which is 2,342,784:

```
mean_series.std()
```

```
2342784.041413037
```

This shows us the standard deviation of the distribution of sampling means. Statisticians call this the standard error of the mean. This chewy phrase would have been clearer, although longer, if it had been something like this: "the standard deviation of the distribution of sample means for samples (each of the same size) drawn from a population." Unfortunately, statisticians are not known for giving things helpful, explanatory labels. Suffice to say that when we are looking at a distribution, and each data point in that distribution is itself a representation of a sample (e.g., a mean), then the standard deviation has the specialized name of standard error.

Second, there is a shortcut to finding out the standard error that does not require actually constructing a distribution of 10,000 (or any other number) of sampling means. It turns out that the standard deviation of the original raw data and the standard error are closely related by some simple algebra:

```
US_region_pops.april10census.std()/math.sqrt(8)
```

```
2391597.645666869
```

The formula in this command takes the standard deviation of the original April 2010 data and divides it by the square root of the sample size. When we created the sampling distribution of means, we used a sample size of n=8. That's what you see in the formula here inside of the sqrt() function. In Python and other programming languages, sqrt() is the abbreviation for square root and not for squirt.

So, if you have a set of observations, and you calculate their standard deviation, you can also calculate the standard error for a distribution of means by dividing by the square root of the sample size. The number we got with the shortcut was slightly different than the number that came from the distribution, but the difference is minor and only arises because of randomness in the sampling distribution.

One last rule of thumb: We previously found the 2.5 percent and 97.5 percent cut points by constructing a sampling distribution and using quantile() to tell us the thresholds. You can also compute cut points using the mean and the standard error. Roughly speaking, two standard errors down from the mean is the 2.5 percent cut point, and two standard errors up from the mean is the 97.5 percent cut point. If you look in the notebook for this chapter, you will notice again that the computed cut points are a little different from earlier. Again, these minor differences arise because of randomness in sampling. The computed values are based on statistical proofs. We could reduce the discrepancy between the two methods by using a larger sample size and/or by having more replications in the sampling distribution.

To summarize, using a population that includes 52 data points from the April 2010 census, and some work using Python to construct a distribution of sampling means, we have learned the following:

- Run a statistical process a large number of times, and you get a consistent pattern of results.

- Taking the means of a large number of samples and plotting them on a histogram shows that the sample means are normally distributed and that the center of the distribution is close to the mean of the original raw data.

- This resulting distribution of sample means can be used as a basis for comparisons. By making cut points at the extreme low and high ends of the distribution, for example, 2.5 percent and 97.5 percent, we have a way of deciding whether a new sample mean is unusual.

- If we get a new sample mean, and we find that it is in the extreme zone defined by our cut points, we can tentatively conclude that the sample that made that mean is a different kind of thing from the samples in the sampling distribution.

- A different way of figuring the cut points involves calculating the standard error based on the standard deviation of the original raw data. This is the more typical method used in statistical formulas.

We're not applied statisticians yet, but the process of reasoning based on sampling distributions is at the heart of inferential statistics, so if you have followed most of the logic presented in this chapter, you have made excellent progress toward being a competent user of applied statistics.

CASE STUDY: ANALYZING THE IMPACT OF A NEW TREATMENT

```
Case Key Points:
 - Did a change in treatment to passengers improve
   the airline's customer evaluations?

 - We must ensure our analysis accounts for expected
   randomness in the results.

 - We can use sampling, replication, and quantile to
   understand if the treatment is better.
```

To improve the airline company's customer ratings at select airports, prior to boarding the flight, free soft drinks were provided to passengers. These customers were then surveyed at the end of their flight. We have access to these surveys and need to figure out if the drinks improved customer evaluations. We need to make sure that any observed difference was not due to sampling error. We can answer this question by comparing the mean LTR value after the treatment with a comparison distribution. It is important to keep in mind that this is an informal method of accomplishing this kind of comparison. A formal statistical approach to making this comparison would require some additional decisions about our data and a corresponding choice of statistical procedure.* With that in mind, let's first read the data of the original survey and the new survey.

```
small_survey = pd.read_csv(\
      "https://raw.githubusercontent.com"\
      "/jmstanto/IDSpython/main"\
      "/smallSurvey.csv")
small_survey = small_survey.dropna(subset=['LTR'])
small_survey = small_survey.reset_index(drop=True)

new_treatment = pd.read_csv(\
      "https://raw.githubusercontent.com"\
      "/jmstanto/IDSpython/main"\
      "/newTreatmentSurvey.csv")
new_treatment.shape

(120, 5)
```

Let's have a peek at the LTR means from both surveys:

```
print(small_survey.LTR.mean())
print(new_treatment.LTR.mean())

7.290581162324649
7.941666666666666
```

These results suggest that the mean from small_survey, which is 7.29, is about two-thirds of a point lower than the mean of 7.94 from the new_treatment survey. That sounds promising, but as we know, the data collected for new_treatment is only a sample, and it is almost certain that it contains some sampling error. Is the mean of 7.94 large enough to be trustworthy—in other words, do we think that the soft drinks made an improvement in customer evaluations over and above natural variation due to sampling error? Now we can create a sampling distribution of means from small_survey and compute the lower (2.5%) and upper (97.5%) thresholds from that distribution:

```
random.seed(1)

sample_means = [ statistics.mean(
```

---

* For the statistically minded, NPS is a characteristic of a whole sample rather than an individual observation. Technically the new treatment gives us only a single observation of NPS to compare to a baseline. We can overcome this by comparing LTR scores between the original survey and the new treatment survey using a t-test. Alternatively, we could characterize each participant in our surveys as a promoter or detractor (and optionally include a third category of neutral for those scoring between 7 and 8). Then we could use a chi-square test of independence on the resulting count data.

```
    random.choices(small_survey.LTR, k=120) )
    for i in range(100000) ]

mean_series = pd.Series(sample_means)
mean_series.quantile([0.025, 0.975])


0.025              6.875000
0.975              7.691667
dtype: float64
```

We have created a large sampling distribution here, with 100,000 means in it, to ensure a precise result. The resulting thresholds suggest that any new sample mean for LTR that is smaller than 6.87 or larger than 7.69 would be unusual indeed in the "untreated" case (i.e., any new sample of passengers who did not get soft drinks). The observed LTR mean from new_treatment was 7.94: That is so large that it would be considered unusual. Therefore we can tentatively conclude that the soft drinks did have a positive effect on customer evaluations. For fun, let's compute the thresholds using the standard error and the mean:

```
std_error = small_survey.LTR.std()/math.sqrt(120)

print(small_survey.LTR.mean() - (std_error * 2))
print(small_survey.LTR.mean() + (std_error * 2))


6.874950665041366
7.706211659607932
```

Those thresholds are essentially identical to what we established by constructing our own sampling distribution, and therefore we draw the same conclusion.

In summary, don't forget that this is an informal demonstration that shows the essential logic of how statistical inference works. Statisticians who worked on the same problem would use a formal procedure called an "independent samples t-test" to perform the comparison. Our demonstration clearly showed the importance, however, of understanding the uncertainty involved in working with samples with data. No individual sample will precisely match the underlying population, so it is important that we keep the uncertainty in mind—and quantify it when we can—when using statistics to make decisions.

CHAPTER CHALLENGES

1. Run the following line of code:
    random.choices(["blue ", "red "], k=8)
    Describe the output that this line of code creates. How many times did your output say blue, and how many times did it say red? Create a fraction with the count

of blue on top and the total number of reds and blues on the bottom. What would you expect this fraction to be?

2. Run the following line of code:

    pd.Series(random.choices(['blue','red'], k=8)) == 'blue'

    Examine the output, and describe what is different about the output of this line of code when compared to the results of the previous question. Python represents values of True as one and False as zero. Enclose that complete line of code with a call to the sum function. Explain the results.

3. Create a custom function using the code you created for the previous problem. Test the function: When you call it, the function should return the number of blues. Replicate that 1,000 times. Store the result, which should be a vector of integers between 0 and 8 in a new variable. Create a histogram of the new variable. Describe the shape of the histogram.

4. Use random.normalvariate() to create a vector of 1,000 random numbers, and store those numbers in a new variable. Create a histogram of the new variable. Describe the shape of the histogram.

5. Using the list of random numbers from the previous problem, use the quantile() function to find the value for the 2.5 percent (0.025) quantile and the 97.5 percent (0.975) quantile. What percentage of your vector of observations falls in between these two numbers?

6. Display the vector of random numbers from the previous problem on a histogram. Use the axvline() function to draw vertical lines on the histogram at the 2.5 percent and 97.5 percent quantiles. If you observed a new value that was equal to 3.0, would it fall in the central region of this distribution or in one of the tails?

7. Create a dataset consisting of at least 20 data points, and construct a sampling distribution. Calculate the standard error, and use this to calculate the 2.5 percent and 97.5 percent distribution cut points. The data points you collect should all represent instances of the same phenomenon. For instance, you could collect the prices of 20 textbooks or count the number of words in each of 20 paragraphs.

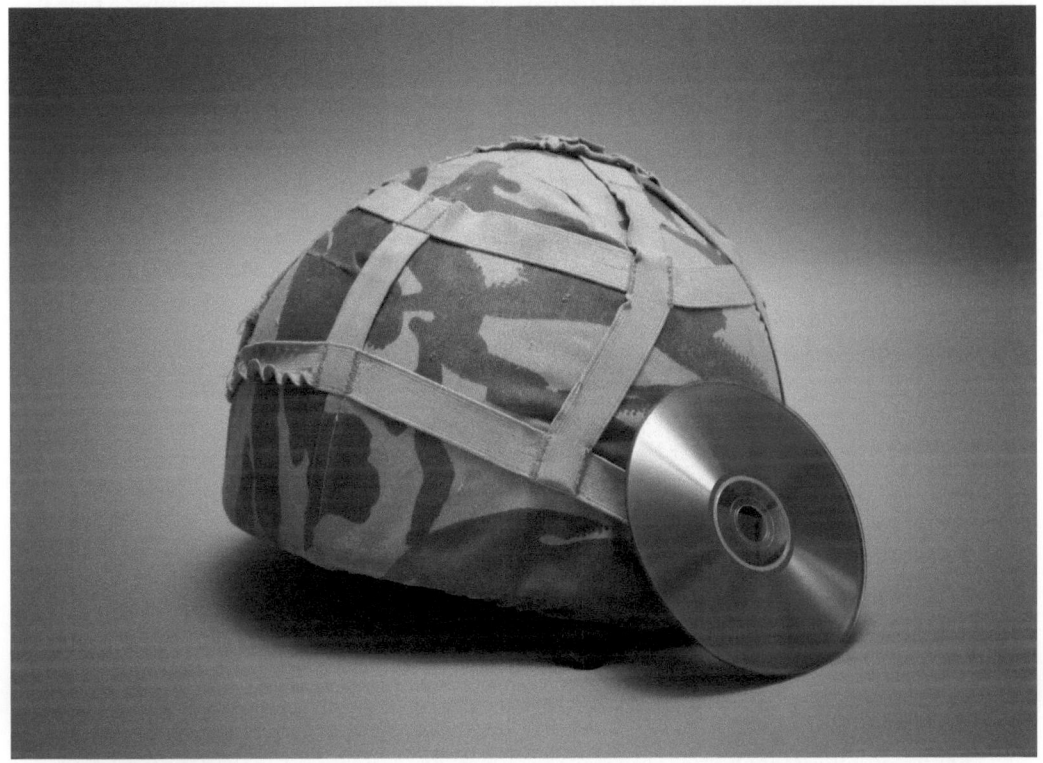

7 STORAGE WARS

LEARNING OBJECTIVES

Evaluate different data sources that are available for data analysis in the Python environment.

Explain the reasons for using file-based data or database data.

Build Python code to access data that are available in Excel, JSON, and an SQL database.

Use the pandas to_sql function combined with the sqlalchemy library to access a DataFrame as if it were a database.

People who have seen the evolution of technology over recent decades may remember a time when storage was expensive and had to be hoarded like gold. Over the past few years, however, the accelerating trend of Moore's Law has made data storage almost "too cheap to meter" (as they used to predict about nuclear power). Although this opens many opportunities, it also means that people keep data around for a long time, because it doesn't make sense to delete anything, and they might keep data around in many different formats. As a result, the world is full of different data formats, some of which are proprietary—designed and owned by a single company such as SAS—and some of which are open, such as the lowly but useful CSV format.

In fact, one of the basic dividing lines in data formats is whether data are human readable or not. Formats that are not human readable, often called "binary formats," are efficient for how much information they can pack in per kilobyte but are also squirrelly in the sense that you need specialized software to retrieve data in that format. As you might expect, human-readable formats are inefficient from a storage standpoint but easy to diagnose when something goes wrong. For high-volume applications, such as credit card processing, the data that are exchanged among systems are almost universally in binary formats. When datasets are archived for later reuse, for example, in the case of government datasets available to the public, they are usually available in multiple formats, at least one of which is a human-readable format.

Another dividing line is between proprietary and open formats. One of the most common ways of storing and sharing small datasets is as Microsoft Excel spreadsheets. Although this is a proprietary format, owned by Microsoft, it has also become an informal standard. Dozens of different software applications can read the formats Excel supports. On the nonproprietary side, the OpenDocument format is an open format, managed by a standards consortium, that anyone can use. OpenDocument format is based on XML, which stands for extensible markup language. XML is a whole topic in itself, but in brief it is an internationally approved data exchange format designed specifically to work on the internet and is both human and machine readable. XML is managed by the W3C consortium, which is responsible for developing and maintaining the many standards and protocols that support the web.

With many free libraries available, Python offers a wide variety of methods of connecting with external data sources. In this chapter, we will explore how to read in a dataset using three common methods. First, we will read in an Excel file. As you will see, this will be similar to how we read in the CSV format. We will also explore extracting data from a source that is not a file. Many data sources, particularly databases, exist not as a discrete file but rather as an entry point into a database system. The database system provides methods or calls to query data records from the system, but from the perspective of the user (and of Python), the data never take the form of a file. Specifically, we will read a dataset that is available via a database using an industry standard known as "structured query language" (SQL). Finally, in the last part of this chapter, we will access data via Java Script Object Notation (JSON), which is a structured, but human-readable, way of sharing data. JSON is an increasingly common way of sharing data on the web, particularly from web-based application programming interfaces (which we discuss in detail later in this chapter).

ACCESSING EXCEL DATA

Because Excel is such a widely used program for small, informal datasets, we will start with an example here to illustrate both the power and the pitfalls of accessing binary data with Python. The easiest way to read in an Excel file is to use pandas, as we did for reading a CSV file. A review of the documentation reveals that the only required argument to this function is the location of the XLS file and that this location can be a pathname. So, let's give it a try.

If you think back to an early chapter in this book, you may remember that we accessed some census data that had population counts for all the U.S. states. Previously, we read the data in a CSV format. In the following cell, we instead read the Excel file containing that data directly into a DataFrame using the read_excel() function:

```
excel_url = ('http://www2.census.gov/programs-surveys/'\
             'popest/tables/2010-2011/state/totals/'\
             'nst-est2011-01.xls')

temporary_df = pd.read_excel(excel_url)

temporary_df.shape

(66,5)
```

The first three lines create the URL of the Excel file we want to download. Next, we import the file with the read_excel() function. The shape of the data seems reasonable, but like when we read in the CSV dataset, it looks like there would be some cleanup, as we had to do for the CSV file. We covered that data cleaning process in Chapter 3 and so will not repeat ourselves here. The notebook file for this chapter contains the read_census() function that we developed in Chapter 4, so all we need to do is use that:

```
US_region_pops = read_census(
    'http://www2.census.gov/programs-surveys'\
    '/popest/tables/2010-2011/state/totals/'\
    'nst-est2011-01.csv')

US_region_pops.iloc[:2,:3]

    state_name    april10census    april10base
0   Alabama       4779736          4779735
1   Alaska        710231           710231
```

Now we are ready to analyze our data or do whatever tasks our project requires. Note that the census files are particularly difficult to work with because of the notes and the unusual column naming. In many cases if you receive a spreadsheet used in a business context, it may already contain sensible column names and data formats that will make the importing process easier. On a related note, we often get the question of whether it is better to manually edit an Excel file to fix it up before trying to import it. The answer to this question depends on the data management processes you are using. On one hand, if the Excel file was automatically generated by some other business process and will be regenerated week after week, or month after month, then you should automate all of the necessary transformations in Python. On the other hand, if someone hands you a spreadsheet for a one-time analysis activity, and the spreadsheet isn't too large, then feel free to do whatever will produce the best result in the shortest amount of time—even if that includes a little manual editing in the spreadsheet.

WORKING WITH DATA FROM EXTERNAL DATABASES

Now we are ready to consider another strategy for getting access to data: querying it from external databases. Depending on your familiarity with computer programming and databases, you might notice that the method of obtaining the data is quite different here. In the previous example, we had a file (sometimes rather messy) that contained a complete copy of the data that we wanted, and we read that file into Python and stored it in our local computer's memory. We also have the option of saving a copy of our Python data object on our local file storage. This is a good and reasonable strategy for small- to medium-sized datasets, which we'll define for the sake of argument as anything up to 100 megabytes.

But what if the data you want to work with is large—too large to represent in your computer's memory all at once and too large to store on your own hard drive? This situation could occur even with smaller datasets if the data owner did not want people making complete copies of their data but rather wanted everyone who was using it to work from one official version of the data. In a similar vein, there are many situations where multiple users need to share access to data: It is much better to have them work with a database that was designed for this purpose. For these reasons, it becomes necessary to do one or both of the following things:

- Allow Python to send messages to the large, remote database, perhaps via the web, asking for summaries, subsets, or samples of the data.
- Allow Python to send computation requests to a distributed data processing system asking for the results of calculations performed on the large, remote database perhaps via a web service.

Like most contemporary programming languages, Python provides several methods for performing these two tasks. We will explore two basic ways to access these remote data services (via a database and JSON).

ACCESSING A DATABASE

The first strategy we will explore to access remote systems of data involves using a library that can connect to an SQL database server. SQL is used for accessing (via queries) data stored in a relational database. Python has many libraries that support SQL. In this situation, Python sends a command to the database server, which then returns a result to the Python client. The library then places it in a Python data object (typically a pandas DataFrame) for use in further processing or visualization.

We will use the open source sqlalchemy library, which is a popular library for working with SQL within Python. First, we import the sqlalchemy library. After that we would normally build a connection to a remote database that was stored on some other server. In this case, we will create our own in-memory database by using the pandas to_sql() command. This procedure will generate a database table for the given DataFrame:

```
import sqlalchemy
from sqlalchemy.sql import text

engine = sqlalchemy.create_engine('sqlite://', echo=False)

US_region_pops.to_sql('census', con=engine)

52
```

In this code, we first import the package and then import a utility function we will need soon called "text()." In the second line of code, we create an in-memory database. The 'sqlite://' option specifies that we will be referring to an in-memory database that follows the structural format of a popular database system called "sqlite3." One of the powerful aspects of sqlalchemy is that it knows how to connect to a wide variety of the most powerful database engines. If we had been connecting to an external database on some other server, this line of code would provide the network or web address of that server as well as any security credentials that we would need to provide to access that external database.

The echo=False argument suppresses the large stream of messages that sqlalchemy normally prints to assist programmers with debugging. You can change this to echo=True to get an overview of the operations that sqlalchemy does to create the database. This line of code produces a Python object called "engine" that provides an access point for working with the sqlalchemy/sqlite in-memory database that we created. The engine object is an instance of a specialized class provided by sqlalchemy as a handle for programmers to use to maintain access to the database in question. Finally, the last line uses our engine object in a call to a pandas method known as "to_sql()." This method instructs pandas to send sqlalchemy both the structure of the pandas DataFrame (this mainly comprises the names of the columns) and the data that pandas already has stored in the DataFrame. The first argument also tells sqlalchemy to name the resulting table "census."

At this point, sqlachemy has created a new data table internally to Python's memory and has copied the pandas data into that space. Going forward the only way to access that database is by means of an SQL connection. The connection is like a two-way pipe that we can send queries into to receive information back. Let's use the engine object to create that connection pipe:

```
connection = engine.connect()

type(connection)

sqlalchemy.engine.base.Connection
```

The first line of code creates the connection by calling the connect() method on our database engine. The second line asks Python to report the data type of the resulting connection object. We can now use that connection object to issue queries and receive information from the database:

```
tmp = connection.execute(\
      text("SELECT * FROM census")).fetchall()

tmp[0] #show the first row


(0, 'Alabama', 4779736, 4779735, 4785401, 4802740)
```

The first line of code calls the execute() method that the connection object makes available. "Execute" is old terminology from the early days of computing that basically means "run these commands." In those early days, everything done with computers was accomplished by issuing text-based commands, and the old computers would not do anything until the user said "execute." The command that we are sending to the database appears on the next line enclosed in a utility function called text() that we imported from sqlalchemy earlier. The execute() method is expecting to receive a specific type of data object that contains the SQL query, and the text() function converts our simple text string into that data object.

The SQL query itself is the most interesting part of this code cell. The command says "SELECT * FROM census." First, note that the SQL keywords SELECT and FROM are capitalized. This is not required by most databases but is a kind of informal convention among database programmers. Some people don't like it because it looks like you are SHOUTING the SQL command, but the choice of whether to capitalize is up to you. When we ask to "SELECT *" we are asking to get all of the columns (the * is a wild card), and when we say "FROM" we are specifying the data table from which the data should be extracted. In this case, "FROM census" means that the data should be taken from the census table. This might seem confusing: Our original pandas DataFrame had the Python name of US_region_pops, but when we used that

DataFrame to set up our in-memory database, we specified that we wanted the resulting database table to be called "census." As a result, all of our SQL calls need to use that name.

Now let's try a more advanced query. The following SQL command hints at the power that can be wielded with knowledge of SQL keywords and syntax:

```
connection.execute(
    text("SELECT state_name,\
    july11pop FROM \
    census WHERE july11pop<1000000"))\
    .fetchall()

[('Alaska', 722718),
 ('Delaware', 907135),
 ('District of Columbia', 617996),
 ('Montana', 998199),
 ('North Dakota', 683932),
 ('South Dakota', 824082),
 ('Vermont', 626431),
 ('Wyoming', 568158)]
```

As before, we are using the SELECT keyword to obtain data. This time, our query asks for two particular columns to be returned, namely, the state_name and july11pop columns. As before, we want the data to come from the census table. At the end of our query, however, we also have a filtering command: "WHERE july11pop<1000000." Given your knowledge of conditionals, you can probably guess what this will do. It will restrict the set of rows that are returned to only those rows that meet the filtering criterion.

By the way, there is a curious aspect to the returned data: it is a list of tuples. You can tell it is a list by the square brackets that start and end the output. In Python, a "tuple" is an arbitrary collection of data elements, and it is expressed as a set of fields separated by commas and enclosed in parentheses. For example, ('Alaska', 722718) is a tuple consisting of two pieces of information: a text string and an integer.

SQL is a highly capable programming language in its own right, and we have only scratched the surface of what it can do. For example, here are two computations that are accomplished on complete columns of data:

```
connection.execute(
    text("SELECT AVG(\
    april10base)\
    FROM census")).fetchall()

[(6009063.980769231,)]
```

```
connection.execute(
    text("SELECT AVG(\
    july11pop - april10base)\
        FROM census")).fetchall()

[(54370.769230769234,)]
```

In the first example, we ask SQL to compute the average of all of the numeric values in the april10base column. You may recognize this number because we used it many times when we worked on sampling in Chapter 6. In the second example, we ask SQL to compute the differences between values in two columns and then take the average of that new set of numbers. The output of the second cell shows that on average states added about 54,731 people between April 2010 and July 2011.

In these code cells we did not assign the results of connection.execute() to another data object, so the results were echoed to the console. But as you saw in an earlier cell, where we requested all of the records from the census table, it is easy to assign the results to a Python list and then perform additional computations on the list. If we wanted the results to be returned as a pandas DataFrame instead of a list, we can ask pandas to issue the SQL call to sqlalchemy and return the result to us as a DataFrame:

```
query = text("SELECT state_name, july11pop FROM \
    census WHERE july11pop<1000000")

df = pd.read_sql_query(query, con=connection)

df

            state_name          july11pop
0   Alaska                      722718
1   Delaware                    907135
2   District of Columbia        617996
3   Montana                     998199
4   North Dakota                683932
5   South Dakota                824082
6   Vermont                     626431
7   Wyoming                     568158
```

That code now brings us full circle: We started by dumping a pandas DataFrame into an in-memory sqlite database. Then we ran a variety of SQL queries on our in-memory database. Finally, we fetched some data from the database and stuck it into a new pandas DataFrame. To finish with this SQL example, let's disconnect from the database system so that we don't leave any loose ends lying around:

```
connection.close()
engine.dispose()
```

The first line closes the connection to the database, although the data table still exists in memory after that line of code runs. The second line of code disposes of the sqlalchemy engine object. Any space used in Python's memory for data tables would then be released. Note, however, that if we had connected to an external database on some other server, that database would still exist, and we could always reestablish a connection to it in the future.

That last point highlights the idea that the typical motivation for accessing data through SQL or another database system is that a large database exists on a remote server. Rather than maintaining our own file copy of the data, thereby having to read a complete dataset into Python's memory of our own complete copy of those data, we can create an engine and then query the SQL database. SQL queries can specify subsets of the data to return to us, can preprocess the data with sorting and other operations, and can compute summaries of the data. SQL is also particularly well suited to joining data from multiple tables to make new combinations. In the present example, we used only one table, it was a small table, and we had created it ourselves in Python from an Excel source, so none of these were good motivations for storing our data in SQL, but this was only a demonstration. Eventually, you may become involved in projects that involve SQL coding, and you may have to reuse SQL code that others have produced. In those cases, using sqlalchemy or another database library to run SQL commands within Python can give you all of the flexibility and power that SQL offers.

ACCESSING JSON DATA

The next strategy we will explore uses an application programming interface (API) to communicate with another application or database system. An API is essentially a specification for how to use a server portal including the address of that portal: When a Python program uses an API, it connects to a server system over a network and controls that server or requests information from it using a set of commands.

In this section, we will explore using JSON to communicate through APIs. JSON is a popular format for sending and receiving data over the web. JSON is a structured, but human-readable, way of representing a hierarchically organized dataset. Although a JSON dataset can be stored in a file, like a CSV or Excel file, a more common use is for a website to use JSON to supply up-to-the-minute information. Although originally derived from the JavaScript scripting language, JSON is language independent and is available for almost all programming languages, including Python!

We will start by exploring how to get currency exchange rate information. Because this information changes frequently, using an Excel spreadsheet of exchange rates can cause problems because the rates stored in a spreadsheet quickly go out of date. We might also want the exchange rate for a specific day in the past. Let's use a website that offers a JSON interface to exchange rate data. There are many websites that offer this capability (you can

Google "foreign exchange rates JSON" to see many possible websites that also provide this data). We will we use openexchangerates.org within our code. Like most of the websites for getting exchange rates, openexchangerates.org requires that we have credentials to do a JSON query. Websites use this strategy so that, if a person or business wanted to make many queries, the website could keep track of and charge for that service. We will not be using that many JSON requests and so can easily register at their website to get a free app ID key. It is a simple process: First you create a free account at https://openexchangerates.org/signup/free. After you get an account, the next step is to obtain an app ID, which we will use on our JSON requests. Note that the free level provides up to 1,000 JSON queries per month.

The open exchange rates API allows a user (program or person) to supply a request. The site will return a list of exchange rates. You can use a different request to get the exchange rates for a specific day in the past. You can access the API over the web using what is called an HTTP GET request. HTTP is the hypertext transfer protocol, and it is the standard method for requesting and receiving web page data. A GET request consists of information that is included in the URL string to specify some details about the information we are hoping to get back from the request. Here is an example GET request to the open exchange rates API:

https://openexchangerates.org/api/latest.json? app_id=yourKeyGoesHere

You can type this request into a web browser as a web address; however, you need to update the "yourKeyGoesHere" with the key you got from openexchangerates.org. The first part of the web address should look familiar: the http://openexchangerates.org part of the URL specifies the domain name like a regular web page. The next part of the URL, api/latest.json? app_id=, tells openexchangerates which API we want to use, and the json indicates that we would like to receive our result in Java Script Object Notation. Finally, the last part specifies our app_id.

As previously mentioned, you can type that whole URL into the address field of any web browser, and you should get a sensible result back. The JSON notation is designed for efficiency rather than readability, but you will see that it makes sense and provides the names of individual data items along with their values. Here's a small excerpt that shows the key elements of the data object that we are trying to retrieve:

```
{
    "disclaimer": "Usage subject to terms…
    "license": "https://openexchangerates.org/license",
    "timestamp": 1590159600,
    "base": "USD",
    "rates": {
      "AED": 3.6732,
      "AFN": 76.699995,
      "ALL": 113.65,
```

You can see that this is a chunk of plain text that has the names of fields in quotes, such as "timestamp," and then each field has a corresponding value. We can use the Python json library to extract the data we need from the structure without having to parse the JSON string ourselves. First, let's create a new helper function to take the address field and turn it into the URL that we need:

```
def compose_rate_URL(my_app_id):
    rURL = "https://openexchangerates.org/api/latest"
    secondPart = ".json?app_id="
    url = rURL + secondPart + my_app_id
    return url
```

There are three simple steps here. After the line with the def statement, the next line places the beginning part of the URL into a string called "rURL." We next initialize the second part of the URL and then combine the separate parts with my_app_id. This creates a complete URL string that you could type into a browser. Let's test our new function with a fake app ID:

```
url = compose_rate_URL('id1234')
url


https://openexchangerates.org/api/latest.json?app_id=is1324
```

Looks good! You can see that this looks like a regular URL that you would type into the address bar of your browser but that it also contains the fake app ID that we provided. If we typed this into a browser, we would get an error message from the website because "id1234" is not a valid app ID. But now that we know the URL builder works, we can use our new function to actually request the data. Pandas has a function call that handles all of the details for us and returns a DataFrame with the results of the API request:

```
appID = 'a4e249a7b00241e2ab5dda91e2999a86'

#get the URL
url = compose_rate_URL(appID)

try:
    fx_df = pd.read_json(url)
    print(fx_df['rates']['EUR'])
except:
    print("Get an appID from the openexchangerates site to get this \
 code block to run.")

0.91805
```

Before you run this code, make sure to go to the website and get your own free app ID. Use that value in place of string for appID in the code. The first thing this code does is to pass our string to compose_rate_URL(). Then the code uses the pandas read_json() function to retrieve the JSON file from the URL, parse the JSON file, and store the results in a DataFrame. From the DataFrame, we can extract the Euro conversion rate. When we ran this example, the exchange rate was 0.91805, which means that one U.S. dollar would buy 0.92 Euros. If you run the example in the notebook for this chapter, you will probably get a different rate. This is why an API is valuable: When information changes dynamically, we can use an API to get the most recent update.

Now we can use JSON for the other important use case: reading a structured dataset. Our next example that we will parse is a dataset about the Citi Bike program in New York City. There are similar programs in many cities. The basic idea is that a person can rent a bike from one bike station and ride it to another station in the city. For example, maybe you take a bike to work, lock it at a station near work, but in the evening, if it's raining, you decide to take the train home. The next day, you can ride a different bike to work. Of course, if it rained every afternoon for a week, and everyone acted the same way, eventually, there would be no bikes at some stations and too many at others. Citi Bike makes data available about how many bikes and spaces are available at each of its stations. One of the ways Citi Bike makes these data available is via JSON. To start, we need to obtain the list of API entry points to locate the URL we need to get available bike information:

```
feed_list_URL = "http://gbfs.citibikenyc.com/gbfs/gbfs.json"

results = pd.read_json(feed_list_URL)

station_info_url = ''

for entry in results.data.en['feeds']:
  if entry['name'] == 'station_status':
     station_info_url = entry['url']

station_info_url

https://gbfs.citibikenyc.com/gbfs/en/station_status.json
```

In response to the call to the API, we have received a JSON structure that pandas has conveniently organized into a DataFrame for us. The DataFrame contains a column called "data," which in turn holds a structure with three language options: English, Spanish, and French. After choosing English, we run a "for loop" that allows us to search for a feed that has the name "station_status." After the loop we print the name of the feed that will provide us with status on each station. So far all we have done is to find the correct URL to call to get station information; next we will actually request that information:

```
results = pd.read_json(station_info_url)

results.columns

Index(['data', 'last_updated', 'ttl'], dtype='object')
```

Pandas received the JSON from this URL and formatted it into a DataFrame with three columns. The 'data' column is the one we are interested in—the other two contain timing information so that we can judge how fresh the data are. The data column contains a list of dictionaries, where each dictionary contains fields that we can examine for useful information. A careful inspection of the whole data structure reveals that the first few stations are not active—probably used for testing—so let's skip ahead to the tenth list entry:

```
results.data[0][10]

{'is_renting': 1,
 'is_installed': 1,
 'num_bikes_available': 7,
 'eightd_has_available_keys': False,
 'num_docks_disabled': 0,
 'station_status': 'active',
 'station_id': 'b4b5824b-3c23-44b7-92a3-cec6a54583a0',
 'num_scooters_available': 0,
 'is_returning': 1,
 'num_docks_available': 14,
 'num_scooters_unavailable': 0,
 'num_ebikes_available': 0,
 'legacy_id': '4169',
 'num_bikes_disabled': 0,
 'last_reported': 1688654954}
```

You can recognize that this is a Python dictionary object by the curly braces at the beginning and the end. We can query any of these fields to find one particular piece of information, such as the number of bikes available:

```
results.data[0][10]['num_bikes_available']

7
```

Given the capability of finding the number of available bikes from any one station, we could tally availability at all of the stations and do some statistics on the results:

```
num_stations = len(results.data[0])
print("There are", num_stations, "stations in total.")

bike_avail_list = []

for i in range(num_stations):
  this_station = results.data[0][i]
  bike_avail_list.append(this_station['num_bikes_available'])

print("Across all stations there are", \
         sum(bike_avail_list), "bikes available.")
print("Average bikes per station:", \
         sum(bike_avail_list)/num_stations)

There are 1956 stations in total.
Across all stations there are 27442 bikes available.
Average bikes per station: 14.02965235173824
```

We could also have plotted a histogram to get the big-picture view of the number of free bikes per station. Also, to provide more accurate results, we should have filtered out inactive stations. Even so, with a few lines of code we have obtained up-to-the-minute information on the status of the Citi Bike system. To accomplish this, we used the capability of querying JSON data from an API entry point. Thanks to pandas, this part of the job is quite straightforward, although finding the data we need within nested data structures still requires some exploration. As a data scientist, one of your roles is to make sense of complex data structures like these and to write code that can extract the key pieces of information needed to make good decisions.

CASE STUDY: READING, CLEANING, AND EXPLORING A SURVEY DATASET

```
Case Key Points:
   - Read a JSON file that has more complete survey data
   - Clean the dataset (survey) by removing NAs and determine
     the percentage of the datafile was removed via cleaning
   - Create an attribute that will have a True/False based on
     if the arrival time was delayed by more than 5 minutes.
   - Create a different attribute that will have a True/False
     based on if that person were a detractor.
```

Let's explore a more complete airline survey dataset that is available via JSON. First, we can read in the new JSON datafile.

```
survey_file = "https://zenodo.org/" + \
              "record/6366242/files/" + \
              "completeSurvey.json?download=1"

survey_with_NA = pd.read_json(survey_file)
survey_with_NA.shape

(88100, 32)
```

This dataset uses about 29 MB of disk space: Depending on where you run this code, the first line might take a minute or two to run. For use in Google Colab, we have hosted the data on a public server that supports large files. When it is done you can see that there are 88,100 rows and 32 columns (attributes). Here's a description of these attributes:

1. **Likelihood to Recommend**—the airline is rated on a scale of 1 to 10, which shows how likely the customer is to recommend the airline to their friends (10 is very likely, and 1 is not very likely).
2. **Airline Status**—each customer has a different type of airline status or package, which are platinum, gold, silver, and blue.
3. **Age**—this is the specific customer's age, from 15 to 85 years old.
4. **Gender**—this is a binary variable encoded as "male" or "female." Note that newer datasets may encode additional options.
5. **Price Sensitivity**—This is the grade to which the price affects the customers purchasing. The price sensitivity has a range from 0 to 5.
6. **Year of First Flight**—This is the first flight of each single customer. The range of year of the first flight for each customer is from 2003 to 2012.
7. **Flights Per Year**—This is the number of flights that each customer has taken in the most recent 12 months. The range is from 0 to 100.
8. **Loyalty**—This is an index of loyalty ranging from –1 to 1 that reflects the proportion of flights taken on other airlines versus flights taken on this airline. A higher index means more loyalty.
9. **Type of Travel**—This describes three traveling purpose for each consumer, which are business travel, mileage tickets based on a loyalty card, and personal travel, for example, to see the family or go on vacation.
10. **Total Frequent Flyer Accounts**—This is how many frequent flyer accounts the customer has.
11. **Shopping Amount at Airport**—This describes the spending in dollars on nonfood or nondrink goods and services at the airport(s) where the customer was before, between, or after flights.
12. **Eating and Drinking at Airport**—This describes the spending in dollars on food and drink goods and services at the airport(s) where the customer was before, between, or after flights.
13. **Class**—This consists of three different kinds of service levels such as business, economy plus, and economy. Moreover, customers have the option to choose their seat.
14. **Day of Month**—This means the traveling day of each costumer. This attribute shows total of 31 days of the month.
15. **Flight date**—All of these data are abbreviate the passenger's flight date travel, which were since 2014 and only in January, February, and March.

16. **Partner Code**—This airline works with wholly and partially owned subsidiary companies to deliver regional flights, for example, AA, AS, B6, and DL.
17. **Partner Name**—These are the full names of the subsidiary airline companies. Pseudonyms have been substituted in place of the real names.
18. **Origin City**—This refers to actual city that customers have departed from, for example, Yuma, AZ; Waco, TX; and Toledo, OH.
19. **Origin State**—This is the same thing as origin city but rather the state that customers have departed from, for example, Texas, Ohio, Alaska, and Utah.
20. **Destination City**—This is the place to which a passenger travels, for example, Akron, OH; Alpena, MI; Austin, TX; and Boston, MA.
21. **Destination State**—This is the same thing as origin city but refers to the state passenger travel to. Some example of destination states are Alaska, Kentucky, Iowa, and Florida.
22. **Scheduled Departure Hour**—This is the specific time at which passengers are scheduled to depart. In this data the scheduled departure hour ranges from 1 a.m. to 23 p.m.
23. **Departure Delay in Minutes**—These are minutes of departure delay for each passenger when compared to schedule. In this data the rage is from 0 to 1,128 minutes.
24. **Arrival Delay in Minutes**—This is the number of minutes arrival is delayed for each passenger. The range of delayed minutes in this data are from 0 until 1,115 minutes.
25. **Flight Cancelled**—This occurs when the airline does not operate the flight at all for any reason.
26. **Flight Time in Minutes**—This indicates the period time to the destination.
27. **Flight Distance**—This is the extent of space between two places and how many minutes passenger travel between two places. The range in this data is from 31 to 4,983 minutes.
28. **olong**—This is the longitude of the origin city.
29. **olat**—This is the latitude of the origin city.
30. **dlong**—This is the longitude of the destination city.
31. **dlat**—This is the latitude of the destination city.
32. **freeText**—This is the free form text response provided by the customer.

A close inspection will reveal that the dataset has a variety of missing data fields. For now, let's remove the surveys with missing data in the Likelihood.to.recommend column, which only removes four rows of data:

```
survey = survey_with_NA.dropna( \
   subset=['Likelihood.to.recommend']).copy(deep=True)

print(survey.shape)

survey_with_NA.shape[0] - survey.shape[0]

(88096, 32)
4
```

Note that the last line of code uses the first entry in the shape attribute, which contains the count of rows, to compute the difference in the size of the dataset before and after removing missing data fields.

Finally, let's create two new attributes. We will hold off doing extensive analysis until future chapters, but for now, let's explore the impact of a big arrival delay on the proportion of people who are detractors. First, we create a new column attribute 'Big.Delay' column.

```
big_list = [(delay > 5) for delay in
survey['Arrival.Delay.in.Minutes']]

survey['Big.Delay'] = big_list

survey[['Arrival.Delay.in.Minutes', 'Big.Delay']][5:10]

    Arrival.Delay.in.Minutes        Big.Delay
5           0.0                     False
6          79.0                     True
7           2.0                     False
8           0.0                     False
9           0.0                     False
```

Next, we can create a second new column to see if the person was a detractor:

```
survey['Detractor'] = survey['Likelihood.to.recommend'] < 7
survey[['Arrival.Delay.in.Minutes',
    'Big.Delay', 'Detractor']][5:10]

    Arrival.Delay.in.Minutes    Big.Delay       Detractor
5           0.0                 False           True
6          79.0                 True            False
7           2.0                 False           False
8           0.0                 False           False
9           0.0                 False           True
```

Now that we have these two new columns, we can group the detractors by if there was a big delay:

```
grouped = survey[['Big.Delay', 'Detractor']]. \
                groupby('Big.Delay').mean()

grouped.rename(columns={'Detractor': 'prop_detractor'})
```

```
                percent_detractor
Big.Delay
False                    0.238406
True                     0.403394
```

The first line of code accomplishes two things in order: First it groups the overall survey dataset based on whether Big.Delay is True or False. Then for each of those subsets, it computes the mean of the Detractor column. This calculation relies on a little trick: In Python, False is encoded as 0 and True is encoded as 1. Calculating a mean therefore equates to calculating the proportion of Trues relative to the size of the data subset. As an example, imagine we have a list with four entries, like this: [True, True, False, False]. Because these are actually encoded as [1, 1, 0, 0]. the mean would be 0.5, which also signifies that half of the entries in this mini-dataset are Trues. So looking back at the output of the previous cell, we can see that approximately 60 percent of the people who had a large delay were detractors, but only 24 percent of the people who did not have a large delay were detractors. This should, of course, not be shocking—people do not like when their flight is delayed!

CHAPTER CHALLENGES

1. Create an Excel file on your local computer that contains at least two columns and at least 20 rows. Make sure that the first row contains variable names. Copy the file into your Google Colab virtual machine. Read the Excel file into Python. Provide an overview of your data once it has been imported into Python, using shape, and describe.

2. Remove the first 16 rows of the DataFrame you created in step (1). Using Python calculations, how many rows are left?

3. Reuse the code from this chapter's notebook to run the following SQL command on the census data: "SELECT state_name, july11pop FROM census WHERE july11pop>1000000".

4. Reuse the code this chapter's notebook and use SQL to calculate the average of the july11pop column in the census table.

5. Use the describe() method to explore the large case dataset that we imported from JSON at the beginning of the chapter. Find and report the two numeric columns that have many NAs (say, more than 1,000). The describe() method provides a field called "count" that shows the number of non-NA observations. Anytime this differs from the overall row length of the survey, you can conclude that there are missing values.

6. Run the dropna() method on the large case dataset that we imported from JSON, but do not specify any columns with the subset argument (thus, missing data will be considered from all columns). How many observations are left once all of the missing data rows have been removed?

7. Explore the web to find other JSON-encoded datasets. Find one that is interesting to you, and read the JSON dataset into Python. Work with the JSON dataset to make sure it can be easily used by creating a DataFrame with the key information that was read using JSON.

8 PICTURES VERSUS NUMBERS

LEARNING OBJECTIVES

Define visualization, and describe how visualization complements statistical analysis.

Describe key characteristics of an effective visualization.

Use the seaborn visualization library to construct histograms, boxplots, line charts, bar charts, scatter charts, and heatmaps.

Vision is sometimes described as the most powerful sense that humans possess. Vision has the highest bandwidth of all our senses in that it processes information rapidly and in parallel. In addition, the brain has specialized neural circuitry for rapid visual pattern recognition—we can scan an image, quickly recognize anomalies, and remember that image. Naturally occurring scenes and artificially created pictures can contain a substantial amount of information that the human visual system can take in at a glance. Our acute ability to perceive, analyze, and remember color, shapes, and patterns thus makes data visualization an essential tool for data scientists.

In the book *Interactive Data Visualization* (2015, Boca Raton, FL: CRC Press), Matthew Ward, Georges Grinstein, and Daniel Keim define visualization as the "the communication of information using graphical representations." Information visualization is thus the use of visual representations of abstract data. This is distinct from visual models and simulations, where digital techniques are used to create a computational version of a real object. Instead, information visualization is used when there is no well-defined two-dimensional or three-dimensional representation of the data. In this chapter, we will explore the essential aspects of information visualization with basic numeric data. In the next chapter, we will additionally explore data that have a geographic component, that is, coordinates on a map.

A VISUALIZATION OVERVIEW

If we wished to quickly review a large grid of numbers (perhaps an Excel spreadsheet) to locate the largest and smallest numbers, it might take a while, and we might easily make a mistake. To make this task easier, we might somehow enhance the grid of numbers with a visual pattern that would make spotting the minimum and the maximum easier.

Choosing the best method to accomplish this becomes clearer if we keep in mind four possible representation methods of every information visualization. Each of these components is in essence a translation mechanism used to change a number into some visual analog. To help think about these methods, let's imagine a simple set of dots (or other symbols) arrayed on a flat surface (like a whiteboard) with an X-axis along the bottom and a Y-axis at a right angle along the left side. Here are four ways of translating a number into a visual aspect of a display or figure:

- **A.** Location: the position of the symbol relative to the X- and Y-axes
- **B.** Color: the color of each dot or symbol—other than location, color is the perhaps the most commonly used method of encoding or labeling data on a graphic; note that 7–10 percent of men have red-green color blindness, so the use of color must take this into account
- **C.** Size: the size of each symbol presented on the plot
- **D.** Shape and Texture: the shape of the symbol and whether the symbol is an outline, a solid color, or a shaded pattern of some kind—shapes and textures are often used to represent categories or groupings of data

Visual designers also consider two other important issues as they imagine how they will translate numbers into a visual representation. First, annotations, such as legends, axis labels, and tick mark labels, make it possible for the viewer of the visualization to recognize and understand the methods the visualization uses to represent data. Second, although visualizations printed on paper are static—they can't be changed without reprinting—those that are displayed on a computer screen have the possibility of interactivity, such as zooming in and zooming out or animation of changes over time. Although these interactive features don't change the essential methods of mapping a number into a visual representation, they do open additional possibilities for what viewers can glean from a visualization. For example, zooming out hides details but might reveal the big-picture shapes that are shown by a pattern of dots.

Before we start exploring how to use these translation mechanisms to create visualizations, one additional point to remember is that we must focus on making sure that a graphic is easy to understand. A common novice mistake is to create a visualization that contains a lot of information but that is confusing to viewers. For example, it is easy to create a visualization that contains so many variables that the person viewing the visualization experiences information overload. In such cases, it could be more effective to actually show less data in one figure. To address this concern, here's a list of 10 authoring principles that provide useful rules of thumb for creating a visualization.

1. **Simplicity.** The statistician and information designer Edward Tufte is famous for suggesting that you should create the simplest possible graph that adequately conveys the information you want to communicate.

2. **Intuitive Encoding.** Consider the encoding used to connect a numeric value to a graphical feature, and make sure the encoding is intuitive. For example, a larger numeric value should be represented by a larger dot and a smaller value by a smaller dot.

3. **Patterns Versus Details.** Particularly for noninteractive visualizations, choose whether the graphic will try to highlight the big-picture patterns or the fine-level details. It is hard to do both in the same graphic unless it has interactive features like zoom.

4. **Value and Axis Ranges.** Select meaningful ranges for the axes. Anchor axes at x=0 and y=0 unless there is a compelling reason to do otherwise.

5. **Data Transformations.** Transforming data can be useful, particularly when a set of numeric values covers a large range. For example, a natural log transformation can compress the range from one to 1,000 into a seven-point scale. Transformations should be documented for the benefit of viewers with annotations such as axis labels and captions.

6. **Avoid Overplotting.** Use transparency and jitter to represent data where many points are plotted near to or on top of one another. "Jitter" is a technique that moves points slightly and randomly so that the presence of multiple points at one location can be perceived more easily.

7. Connections. Use lines to connect sequential data points. This is particularly important when showing a time series.

8. Create Aggregates. You can reduce clutter by combining data into meaningful groups.

9. Comparison Across Panes. When a graphic consists of more than one set of axes, keep axis ranges identical or as similar as possible.

10. Color Palette. Select an appropriate color palette based on the meaning of the data. For example, gradations of temperature might be represented with shades of blue and red, whereas the density of plant life on a map might be shown with yellow and green.

In addition to these 10 authoring principles, keep in mind one overriding meta-principle that data scientists consider. Visualizations are persuasive technology that can have profound effects on important decisions. Misusing one or more of the 10 principles can easily lead to a figure that hides some undesirable aspect or risk factor in the data or overemphasizes some meaningful benefit. Data scientists have an ethical obligation to present data in an unbiased manner and avoid using visual tricks to mislead the viewer's interpretation or conclusions.

With all of those rules and regulations in mind, we are set to begin working some examples. Earlier in the book we used some basic Python graphics during our initial exploration of datasets. For example, when we wanted to understand the distribution of a set of numbers, we used a frequency histogram. This display of the frequencies of occurrence of values in different ranges provided a view of the data that complemented basic descriptive statistics.

BASIC PLOTS IN PYTHON

We will continue to explore the state-by-state census dataset that we have used in previous chapters. In the notebook for this chapter, we begin by loading our libraries. As in Chapter 6, we load matplotlib, which will allow us to do some basic visualizations. You can see that we also loaded a one new library, called seaborn, which is a visualization library built on top of matplotlib. Why is it called seaborn? The author of seaborn liked the television show known as *The West Wing*, and this package is named after Samuel Norman Seaborn (sns), who is a fictional character portrayed by Rob Lowe on that show. After importing the libraries, we load our census data using the same methods as shown in previous chapters.

Now that we have the dataset, let's begin by reviewing the Python graphics that we have previously used. For example, if we want to understand the distribution of the state populations, we can show the histogram depicted in Figure 8.1.

The code for this is the same as what we used in Chapter 6. Remember that a frequency histogram groups together similar data points in bins, thereby providing a summary of the

FIGURE 8.1 ■ **July 2011 Census Data for U.S. States and Territories**

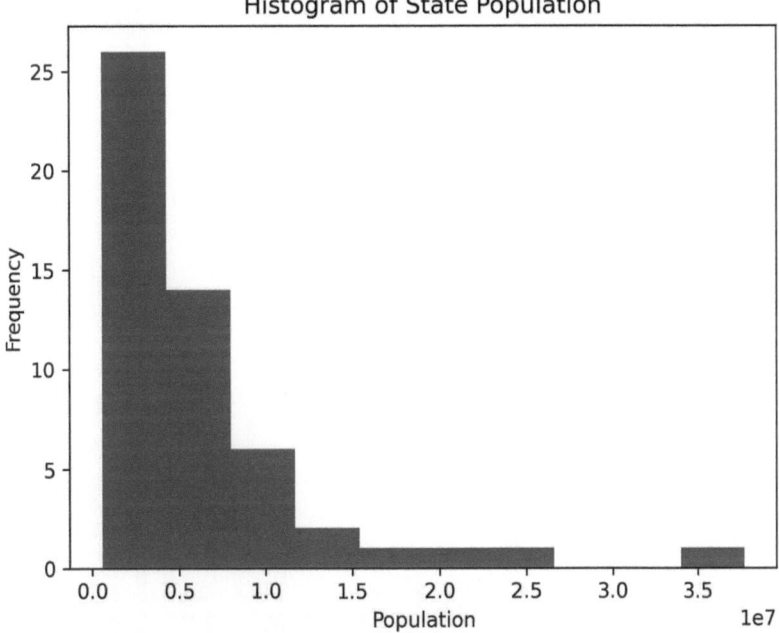

frequency of occurrence of data values at different points on the X-axis. As an alternative to a frequency histogram, we can use matplotlib to show a bar plot (a bar chart) of the actual population values for each state. Figure 8.2 shows this result.

The code to create a bar plot is quite similar to the code for a frequency histogram.:

```
plt.bar(US_region_pops.state_name,US_region_pops.july11pop)
plt.xlabel('stateName')
plt.ylabel('july11pop')
plt.xticks(rotation='vertical', fontsize=8)
plt.savefig('8_2.pdf', bbox_inches='tight')
plt.show()
```

The first line of code specifies the X-axis—which should be categorical for a bar plot—and the Y-axis. In this case the Y-axis is the integer values stored in the column july11pop. We've added axis labels for both the X- and Y-axes. The xticks() function requests that the category labels, that is, the names of the states, be rotated to be vertical. With 52 categories and long names like "District of Columbia," it is not feasible to have the category labels be horizontal.

You might notice that the labels in Figure 8.2 are ordered alphabetically with Alaska first and Wyoming near the end (Puerto Rico is out of order). There could be an advantage to this

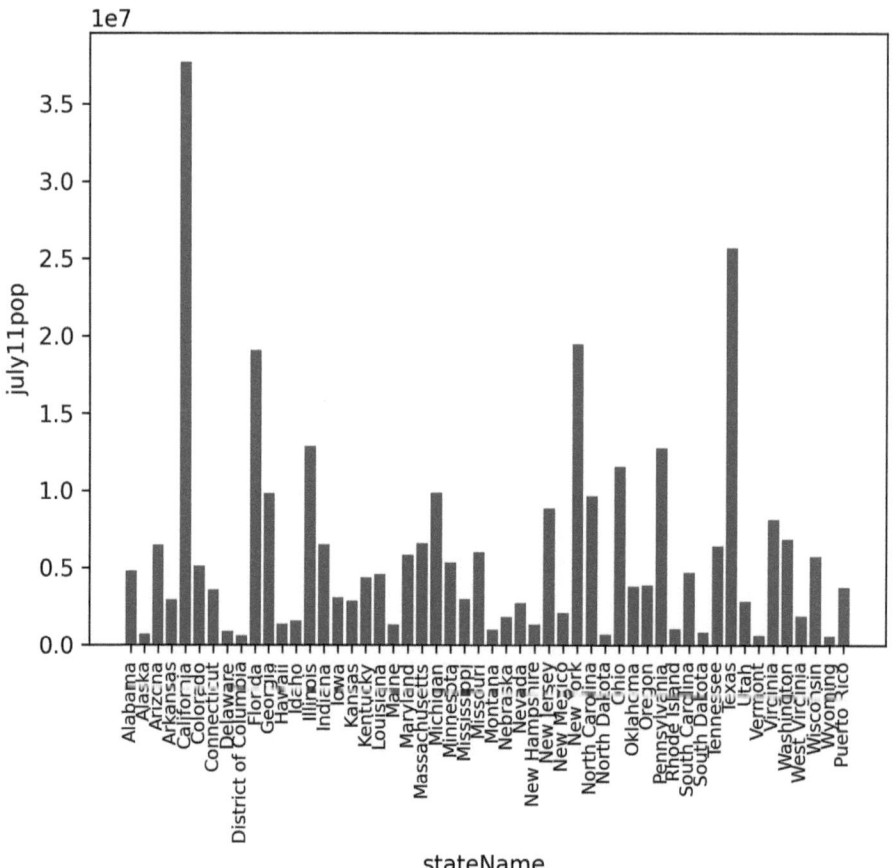

FIGURE 8.2 ■ July 2011 Population Values by State and Territory

presentation if the viewer needed to look up a particular state. From an information design standpoint, though, that task would be better accomplished with a data table. Figure 8.2 does not make clear any particular patterns in the data. For example, when you look at the histogram in Figure 8.1, it is pretty obvious that there are many small states and few large ones. Our bar plot in Figure 8.2 might be more interesting and informative if we sorted the states by population size. In the next section, we will use seaborn to accomplish this.

USING SEABORN

Although plotting these simple charts with matplotlib is easy, if we want to create more-advanced visualizations, we need to explore a more-advanced visualization package. There are actually many different advanced visualization packages that are available, but we will focus on seaborn, which provides a higher-level interface for drawing attractive and informative statistical graphics. Seaborn is perhaps the most popular advanced visualization library for Python, and because it is actually based on matplotlib, if you want, you can use both of those visualization libraries together.

Conceptually, the components of every plot can be imagined as follows:

Geometry: Defines the type of graphic or visualization (histogram, box plot, line plot, density plot, dot plot, etc.). In Python, the type of plot is often selected by choosing the most appropriate function call.

Data: What data elements will be visualized. This is typically accomplished by specifying one or more columns in a DataFrame or more simply by providing a list of numbers.

Attributes: Controls attributes such as the color, size, shape of points, height of bars, and so on.

We can begin our exploration of seaborn by replicating the histogram from Figure 8.1. Here is the code that uses seaborn to create a histogram of july11pop:

```
import seaborn as sns
ax= sns.histplot(US_region_pops.july11pop,
                 kde=False, bins=10)
ax.set(ylabel='Frequency',
       xlabel='Population')
plt.savefig('8_3.pdf', bbox_inches='tight')
plt.show()
```

The first line of code calls seaborn's histplot() function, using the sns abbreviation we defined when we imported the package. The argument kde=False suppresses an additional graphical element that seaborn provides. The abbreviation kde stands for kernel density estimation: In addition to plotting the discrete frequency bins, seaborn can include a smooth, continuous curve that statisticians sometimes use as a diagnostic. The call to histplot() also requests a specific number of bins. We have chosen this value based on our past experience with this distribution because it provides a pleasing representation of the right-hand tail with California as a clear outlier.

We have also set a label for the Y-axis and the X-axis. You could also use this command to include a title for the figure. When creating a figure for a paper or a report, sometimes a figure will have a title and sometimes it will have a caption—less often it will have both. The call to savefig() creates a PDF version of the figure that could be imported into other documents and requests that the bounding box surrounding the figure have as little white space as possible.

You will notice that other than the call to sns.histplot(), all of the other calls are to matplotlib functions. This shows that seaborn is a kind of overlay or enhancement to matplotlib; at heart the figures created by seaborn are matplotlib figures. Figure 8.3 shows the resulting histogram.

FIGURE 8.3 ■ Histogram of July 2011 Population Values

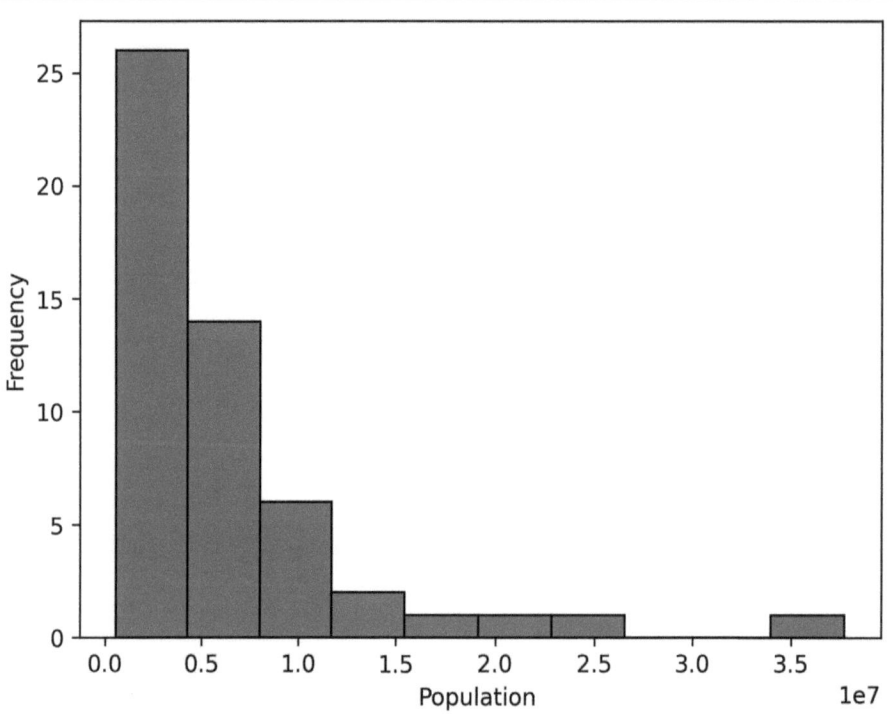

Another way to examine the distribution of a set of numbers is to create a boxplot. Boxplots are an important statistical tool because they provide analysts with the ability to see skewness and outliers at a glance. A boxplot consists of a box, whiskers, and outliers. The box contains the middle 50 percent of the data, whereas the whiskers cover about 99 percent of the data, not counting outliers. The boxplot procedure has an algorithm to decide on what data points get shown as an outlier—often this contains any point that is further away from the box than 1.5 times the width of the box. Figure 8.4 shows the boxplot. Here's the code:

```
plt.figure(figsize=(6,6))
sns.boxplot(y="july11pop",
            data=US_region_pops)
plt.savefig('8_4.pdf', bbox_inches='tight')
plt.show()
```

We know from earlier plotting that our population data are skewed to the right (i.e., a long tail on the high end of the X-axis). That skewness is also evident from the boxplot, which has its lower whisker close to the box, while the upper whisker is much further away. There are also four outliers in Figure 8.4, plotted as little diamonds above the top whisker, that show how far above the norm the four largest states are. We can explore the distribution of the population more in depth by putting the states into two groups—one group for those states

FIGURE 8.4 ■ Boxplot of July 2011 Population Values

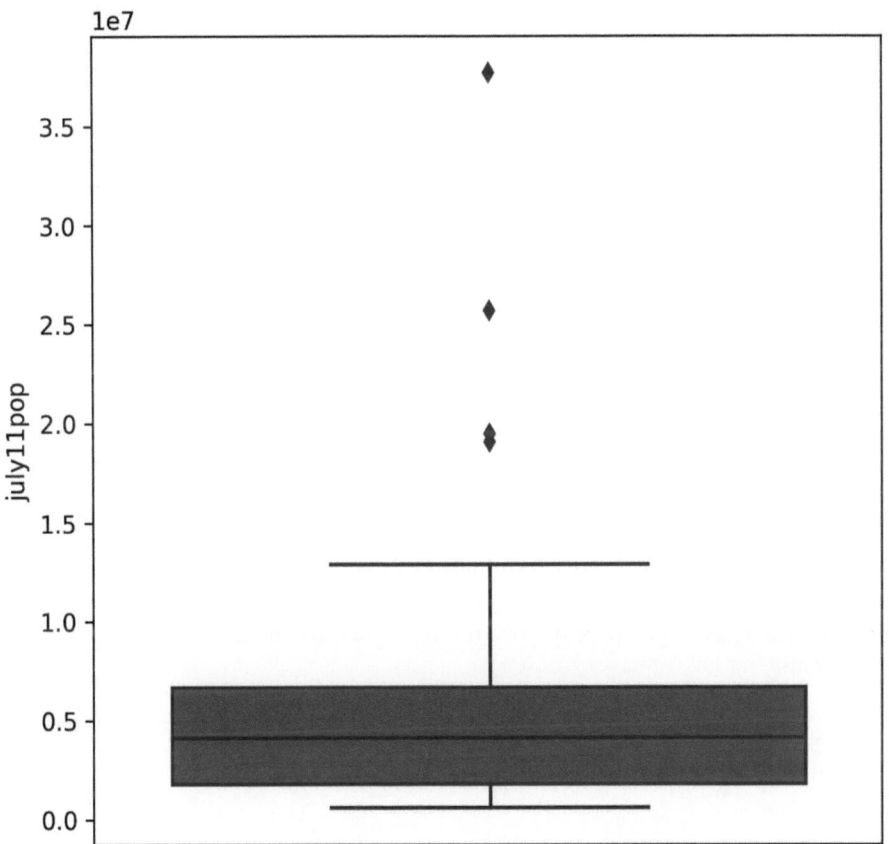

with an increase in population and one group for states with a decrease in population. We can use the following code to create the new variables we will need:

```
US_region_pops['pop_change'] = \
    US_region_pops.july11pop - US_region_pops.july10pop

US_region_pops['percent_change'] = \
    abs(US_region_pops.pop_change/ \
    US_region_pops.july11pop * 100)

US_region_pops['change_direction'] = \
    ['positive' if x > 0 else 'negative' \
    for x in US_region_pops.pop_change]
```

```
US_region_pops_updated = US_region_pops.drop( \
   columns=['april10census', 'april10base', 'july10pop'])

US_region_pops_updated.head(2)

    state_name   july11pop   pop_change   change_direction   percent_change
0      Alabama     4802740        17339           positive         0.362331
1       Alaska      722718         8572           positive         1.186078
```

Note that the first assignment creates a new column in the DataFrame based on the change in population. The second assignment scales that change as a percentage of the july11pop value, using the absolute value of the change, because the third assignment captures whether the population change was positive or negative. We can use this as a grouping variable: states that grew go into the positive group, and states that shrank go in the negative group. Finally, the last assignment drops columns we don't need for now.

With this new DataFrame, we can now create two parallel boxplots, where the variable we are actually plotting is percentage of change, but grouped by increasing and decreasing population, which we can do by defining the x attribute to be the change_direction column, which has the value of positive if a state grew and negative if a state declined in population:

```
palette = {
'positive': 'tab:blue',
'negative': 'white',
}
sns.boxplot(y="percent_change",
x='change_direction',
        data=US_region_pops, palette=palette)

plt.savefig('8_5.pdf', dpi=300, bbox_inches='tight')
plt.savefig('8_5.png', dpi=300, format="png", bbox_inches='tight')
plt.show()
```

Figure 8.5 displays the amount of change for two groups of state, depending upon whether a state grew or shrank in population. You can see in this figure why we took the absolute value of the percentage of change: Both of the boxplots appear on the same Y-axis scale ranging from no change to about 2 percent change. This makes comparing the two groups clearer: It is obvious from the graph that the amount of population loss in the declining states was generally less than the increase in the growing states. This implies that the growing states must have increased their populations through new births and/or immigration because the amount

FIGURE 8.5 ■ Comparing Percentage Change Between Positive and Negative Growth States

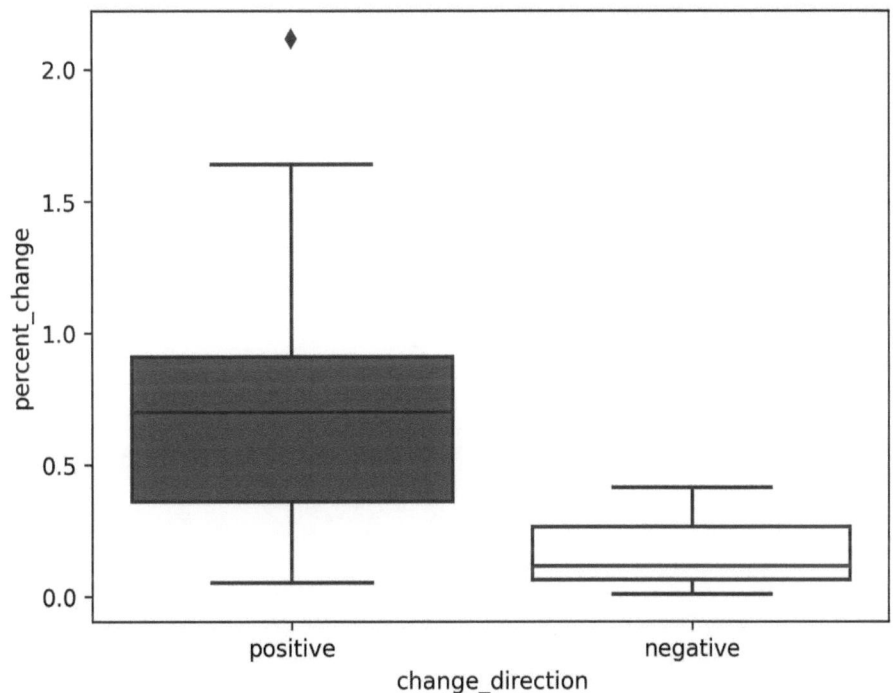

of outflow from declining states does not look like it would match the growth in increasing states.

In addition to histograms and boxplots, we can also use Seaborn to create line charts. To create a line chart, we use the lineplot() function. We rotate the x labels (the state names) so that we can easily read the state name. Finally, note that the height of line represents the value in a column of our DataFrame (the July 2011 population). The code follows, and Figure 8.6 displays the resulting line plot:

```
sorted_df = US_region_pops_updated.sort_values('july11pop')

sns.lineplot(x="state_name", y="july11pop",
             sort=False, data=sorted_df)

plt.xticks(rotation=90, fontsize=8)
plt.savefig('8_6.pdf', bbox_inches='tight')
plt.show()
```

FIGURE 8.6 ■ Line Plot Displaying July 2011 Populations From Smallest to Largest

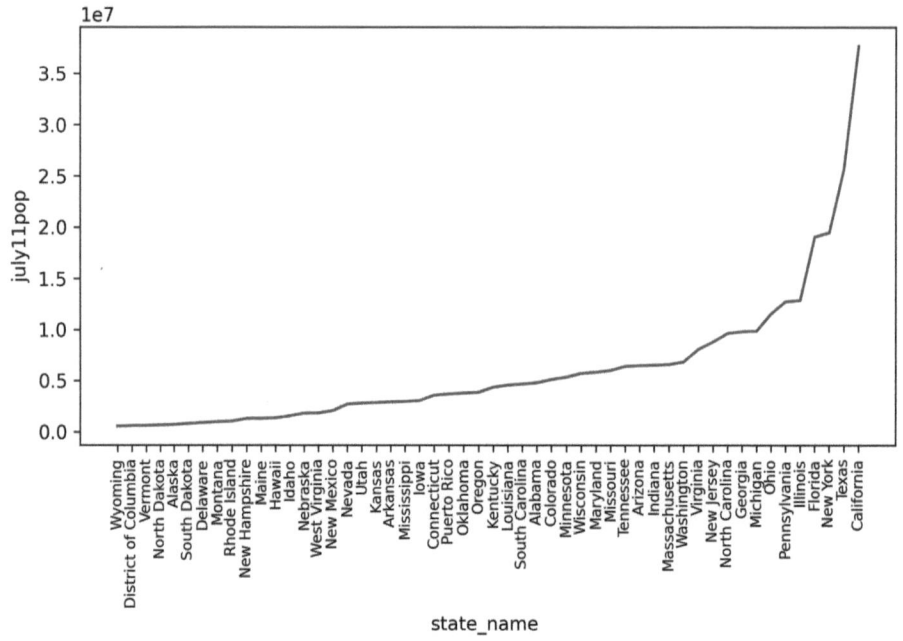

In reviewing Figure 8.6, using a line instead of a sequence of bars reduces the amount of clutter (compare, e.g., with Figure 8.2). Thanks in part to sorting the data before plotting, the pattern becomes obvious: The vast majority of states are relatively small in population, whereas the largest four or five states are considerably larger than any of the others.

One advantage of using a bar chart instead of the line chart is that there is the possibility of adding a third variable into the mix. The following code uses the same sorted data, but we color-code each bar depending upon whether the state grew or lost population.

```
ax = sns.barplot(x='state_name', y='july11pop', \
     data=sorted_df,hue='change_direction', dodge=False)
plt.xticks(rotation=90)
plt.savefig('8_7.pdf', bbox_inches='tight')
plt.show()
```

Figure 8.7 displays all of the population values in the form of a bar plot, but the declining states are shown in a different color than the growing states. If you review the notebook for this chapter, you will see that the former are displayed in orange, whereas the latter are in blue.

FIGURE 8.7 ■ Sorted Bar Plot—Bar Colors Represent Increasing or Decreasing Population

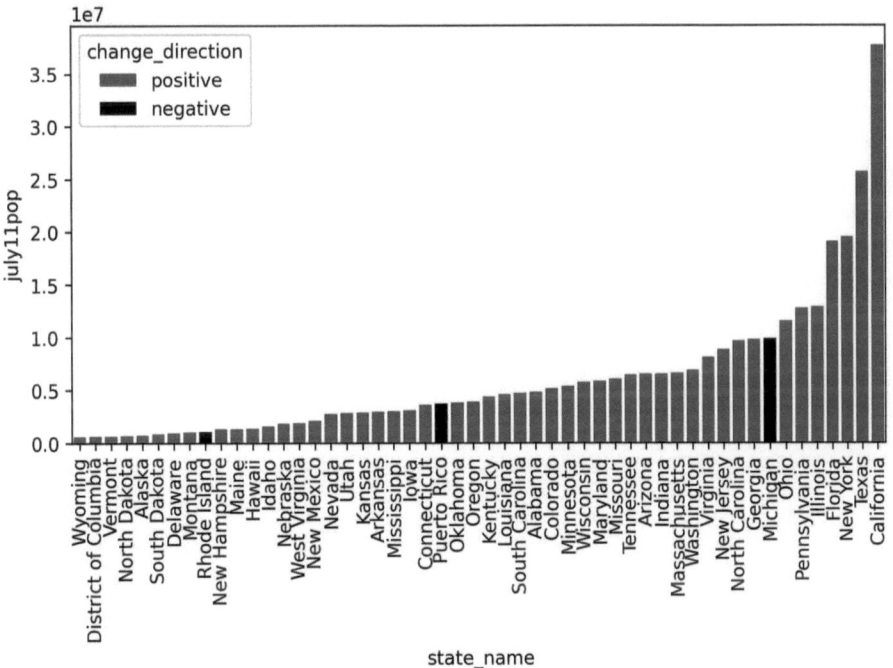

In this book we use black for negative population growth and grey for positive population growth.

SCATTERPLOT VISUALIZATIONS

Another visualization we will explore is a scatterplot, also sometimes called a "scattergram," which provides a way to position data points across two or more dimensions. Figure 8.8. depicts a scatterplot where each point represents a state. Each point shows the raw population change on the X-axis and the percent change on the Y-axis. We represent the direction of change with the color of each dot and the absolute size of the state as the size of the dot. Here's the code to accomplish that:

```
plt.figure(figsize=(8,8))
palette = {
    'positive': 'tab:blue',
    'negative': 'black',
}
ax = sns.scatterplot(x="pop_change", y="percent_change",
                     data=US_region_pops,
```

```
                         hue='change_direction',
                         size = 'july11pop', palette=palette)
plt.savefig('8_8.pdf', bbox_inches='tight')
plt.show()
```

Let's pause and evaluate Figure 8.8. The pattern of dots suggests a mostly linear pattern, where the percentage value of the population change increases as the raw value changes. This pattern would be expected if every state were about the same size. But the pattern breaks with several points in the center and upper right, where the raw value of population change seems large compared with the bulk of the states. We would guess that those three points may include California and Texas, but our graph does not show the state names, so

FIGURE 8.8 ■ A Scatterplot Representing Four Variables

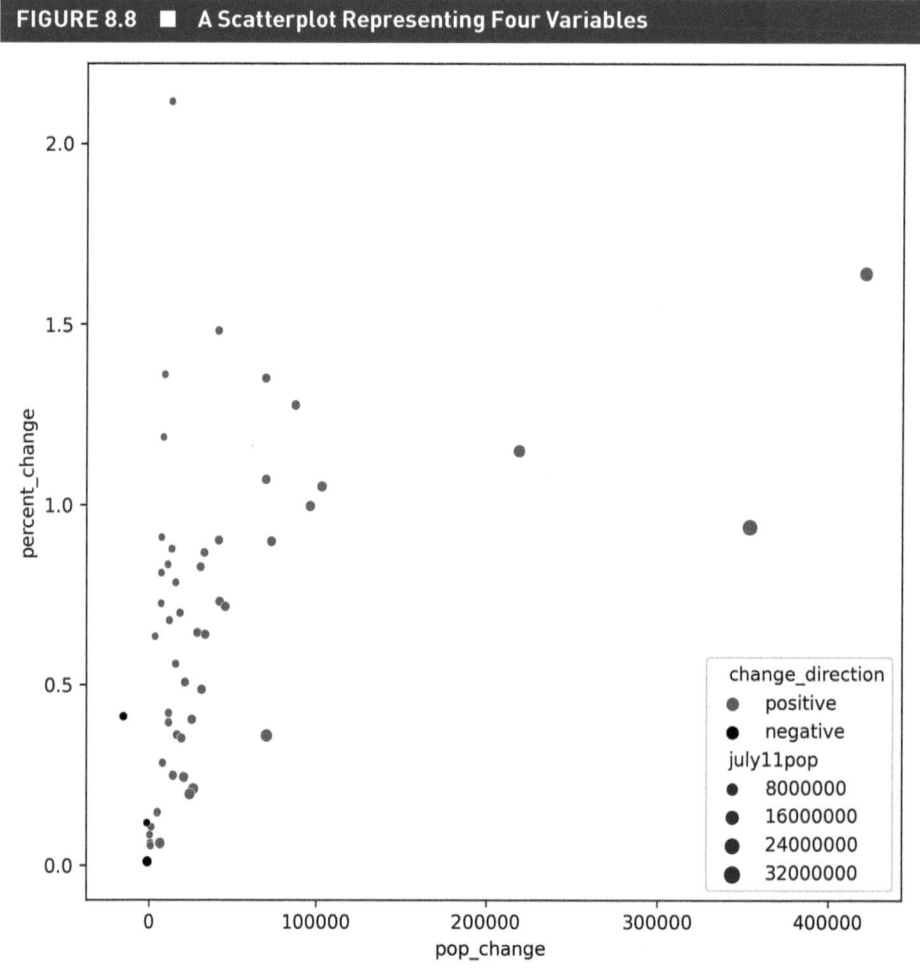

we are not sure. Relatedly, although we might feel a sense of accomplishment for creating a figure that includes the consideration of four different variables, it might be difficult to communicate these results to an audience because of the complexity of the figure. If we were to add labels for the states, these might help tell a more coherent story of population growth.

The notebook for this chapter contains the code and graphical output that shows what happens if we try to label all of the states. Basically, the figure turns into a mess because the lower left corner contains such a concentration of small states that all of the labels overplot each other. Additionally, our storytelling about the data does not get any easier if we label absolutely everything. Instead, let's create a filtering variable that will allow us to focus the audience's attention on a few key states. Here's a code cell that creates a new variable called "key_state:"

```
min_per_change = 1.4
min_pop_change = 100000

US_region_pops['key_state'] = \
    (US_region_pops['pop_change'] > min_pop_change) | \
    (US_region_pops['percent_change'] > min_per_change)

print(US_region_pops['state_name']\
   [US_region_pops['key_state']] )

4                California
8       District of Columbia
9                  Florida
10                 Georgia
43                   Texas
44                    Utah
Name: state_name, dtype: object
```

This code creates two filtering criteria that we then apply to each state to create a True/False map of which states we consider to be key states. We've set the filtering criteria to match the states that have a percentage change of at least 1.4 percent or a population change of at least 100,000 people. There's nothing magic about these choices: You can experiment with changing these criteria to expand or contract the resulting list. Using these criteria, we set up a new column in our DataFrame, key_state, to be True if that state fits our defined criteria of percentage and population change and otherwise False. The code cell then uses this True/False map as a row selector and prints a list of only those states where key_state is True. The output shows that we have listed six different states that have key_state=True, and that list contains California, Texas, and a few others that we had not paid attention to before. Now we can add some labeling to our figure to show the name of each state—but only for the key

FIGURE 8.9 ■ Scatterplot With Key States Labelled

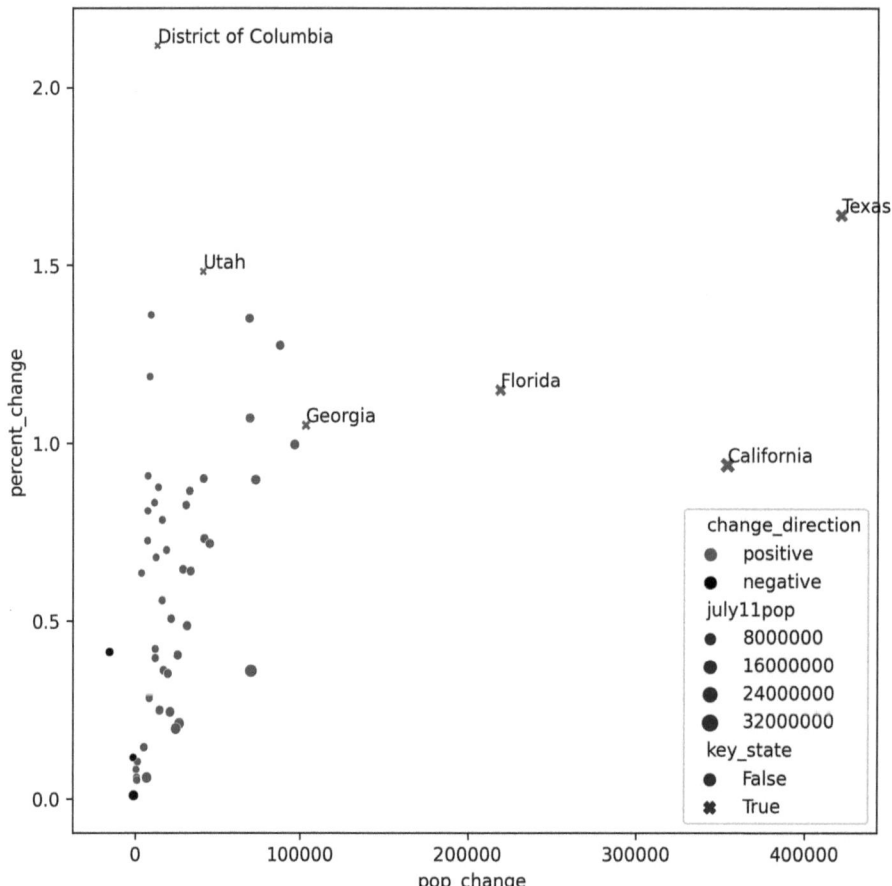

states. Note that we defined some additional aesthetics for the text (the size and the vertical adjustment of where to place the text).

The scatterplot that appears in Figure 8.9 helps tell a coherent story. The vast majority of states are small and have a linear relationship between the absolute value of the population change and the percentage of population that change represents. There are notable exceptions, however, and in three cases—Texas, California, and Florida—we have large states that also had one year of population growth of 1 percent or more.

Examine the following code, which adds state names as labels, but only for the key_states:

```
ax = sns.scatterplot(x='pop_change', y='percent_change', \
                    data=US_region_pops, \
                    hue='change_direction', \
                    size ='july11pop', \
                    style = 'key_state', palette=palette)

states = \
US_region_pops[US_region_pops['key_state']]['state_name']

for state in states:
    x = US_region_pops['pop_change']\
        [US_region_pops['state_name']== state]
    y = US_region_pops['percent_change']\
        [US_region_pops['state_name']== state]

    plt.text(x+0.01, y+0.01, state, fontsize=10)

plt.savefig('8_9.pdf', bbox_inches='tight')
plt.show()
```

Right in the middle of this code cell, there is an assignment statement that creates a temporary variable called "states." By using US_region_pops['key_state'] as a row selector, this statement makes a list of state_names but only for those states that we designated as key states in the previous code cell. We then use that temporary variable to control a for loop that adds the labels to the figure with a call to plt.text().

Also notice how plt.text() includes a slight positive offset for both the X and the Y coordinates. This ensures that the label is not plotted right on top of the point but rather appears slightly up and to the right of the point. If you are used to creating figures automatically with tools such as spreadsheets or presentation software, you might wonder why all of this is not done automatically. The reason lies in the trade-off between convenience and control: Creating graphics with a programming language like Python gives extensive control over all of the visual elements and data mappings, whereas graphing with a spreadsheet is automated but often lacks in customizations.

To conclude this discussion, in Figure 8.9 we have created and labelled a scatterplot that shows us the overall pattern in a set of data but that also highlights the data points of greatest interest because of their population growth, such as Georgia, Florida, and Texas. With this strategy, we have effectively put many of the states with low growth into the background, keeping the level of clutter in our figure as low as possible. To make sure that you understand all of the code in that display, make and test adjustments. For example, increase or decrease the value of min_per_change in the code that creates the list of key_states. In the next chapter we will continue to explore visualizations by expanding our focus to visualize data that can be shown on a map.

CASE STUDY: VISUALIZING KEY ATTRIBUTES RELATED TO NPS

```
Case Key Points:
- Create a function to read our survey dataset

- Visualize different combinations of attributes
and connect attributes to NPS
```

We will continue to use the new survey dataset, so let's create a function to read it into a DataFrame and remove any NAs in the Likelihood to Recommend column. At the same time, we will also add columns for the new variables that we experimented with in the previous chapter.

```python
def read_survey():
    '''
    Reads the airline survey in JSON format from a
    data repository.

    Adds columns for Big.Delay and Detractor. Drops rows
    that have NA in the Likelihood.to.recommend column.

    Arguments: None
    Returns: Pandas Dataframe
    '''
    survey_file = "https://zenodo.org/" + \
        "record/6366242/files/" + \
        "completeSurvey.json?download=1"

    survey_with_NA = pd.read_json(survey_file)
    survey['Big.Delay'] = \
                survey['Arrival.Delay.in.Minutes'] > 5
    survey['Detractor'] = \
                survey['Likelihood.to.recommend'] < 7
    survey = \
            survey.dropna(subset=['Likelihood.to.recommend'])
    print (survey.shape)
    return survey

(88096, 34)
```

This code resembles the steps we took before but has the advantage of being encapsulated in a function. Remember that calling this function will take a few minutes because of the size of the dataset.

A great starting point in getting to know a dataset is to first look at histograms, similar to what we did in Chapter 5. Let's review the overall distribution of the Likelihood. to.recommend attribute. We will use the seaborn histplot() function:

```
ax = sns.histplot(
    survey['Likelihood.to.recommend'],
    kde=False, bins=10)
ax.set(ylabel='frequency')
plt.savefig('8_10.pdf', bbox_inches='tight')
plt.show()
```

This code produces the histogram shown in Figure 8.10. The shape of the distribution suggests that customers are relatively unlikely to use the lower part of the scale.

Note the calibration of the Y-axis in Figure 8.10: We are now working with a large number of observations such that the smallest bin contains about 500 observations, whereas the largest bin contains nearly 20,000. We know, however, that different types of travelers tend to have different opinions about their flights. Let's explore the distribution across the different types of travel using the catplot() function within seaborn to produce Figure 8.11:

FIGURE 8.10 ■ Frequency Histogram of the Likelihood to Recommend Variable

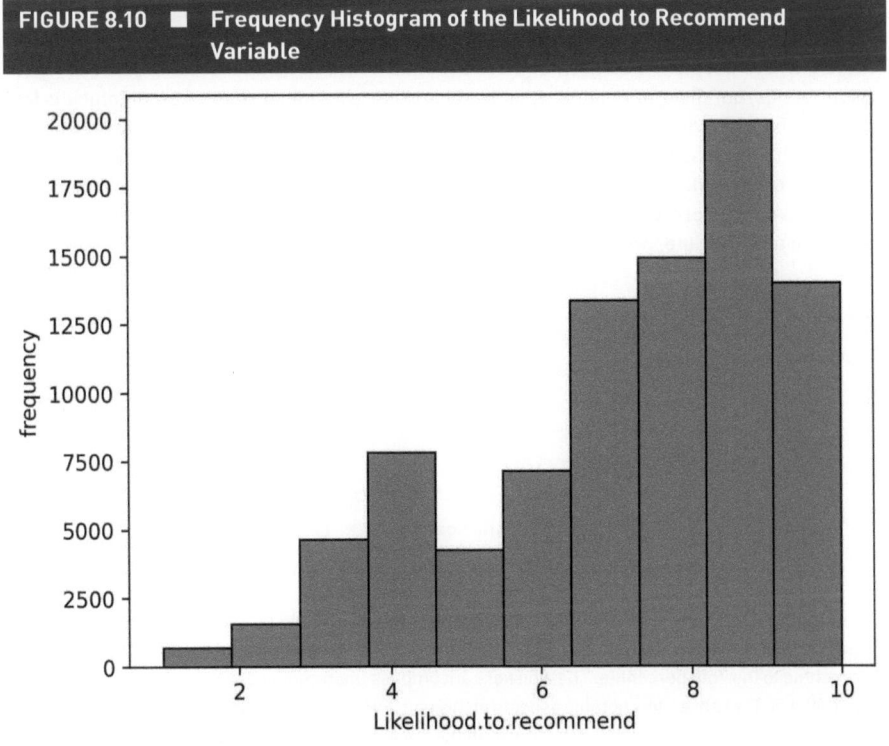

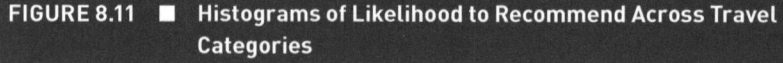

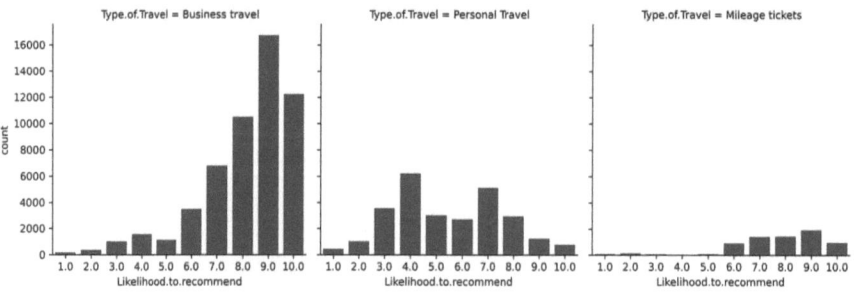

```
ax=sns.catplot(x="Likelihood.to.recommend",
               color="tab:blue", col="Type.of.Travel",
               data=survey, kind="count", height=4)
plt.savefig('8_11.pdf', bbox_inches='tight')
plt.show()
```

Glancing at Figure 8.11, we can see that business travelers outnumber personal travelers and mileage tickets (i.e., when people cash in points for a free flight) by a substantial amount. Also, business travelers tend to provide LTR ratings toward the top end of the scale, whereas ratings from personal travelers have a more symmetric distribution that has two modes: a peak at four with people who are apparently dissatisfied and another peak at seven, where people are somewhat more satisfied. Mileage ticket travelers generally give high ratings, but there are not many of this type. Figure 8.11 also shows how helpful it is for there to be a common Y axis across the three histograms. This graph starts to give some insight into the possible drivers for the positive and negative experiences of fliers.

Let's now examine a different possible connection to ratings, namely, the age of the traveler. As we did before, we will create age groups by rounding to the nearest decade. With this rounded age attribute, we can explore each age group via a boxplot:

```
survey['Rounded.Age'] = survey['Age'].round(-1)

#now visualize the rounded age
ax = sns.boxplot(x='Rounded.Age', \
                 y='Likelihood.to.recommend', \
                 data=survey, palette="Blues")
plt.savefig('8_12.pdf', bbox_inches='tight')
plt.show()
```

Figure 8.12 shows the resulting grouped boxplot with a separate box and whiskers element for each decade of traveler. Remember that the box encompasses half of the distribution from the 25th percentile to the 75th percentile. The whiskers encompass the remaining observations unless there are outliers. There are a few notable aspects of this plot. First, we have used the "Blues" color palette to fill the boxes, though they will appear grey in the book. If you omit the palette="Blues" argument, you will get an even more colorful plot, but the shadings do not convey any attribute of the data except to provide a bit of visual differentiation among the different age groups.

There's one apparent anomaly in Figure 8.12. If you examine the box for the 40-year-old decade, there is apparently no median line. This can occasionally happen with a highly skewed distribution. In the notebook for this chapter, you will see a follow-up code cell that displays the median for this group as equal to nine. That signifies that the median and third quartile are essentially right on top of one another: The box appears to have no median line because it is jammed up against the top of the box. That information, plus the long downward tail and the three outlier dots, indicate that the distribution of LTR for 40-year-olds is strongly negatively skewed. This is confirmed with a bonus histogram that appears in the notebook for this chapter.

Now that we have examined the distribution of LTR, let's visualize some of the statistical properties of the NPS because as it is also an important focus of our overall analysis. One way to create insight into a variable is to produce output for different groups. We saw in Figure 8.12 that sometimes the graphics function itself can produce that grouping, as is true with seaborn's catplot() and boxplot() functions. In other cases, we can produce the grouping

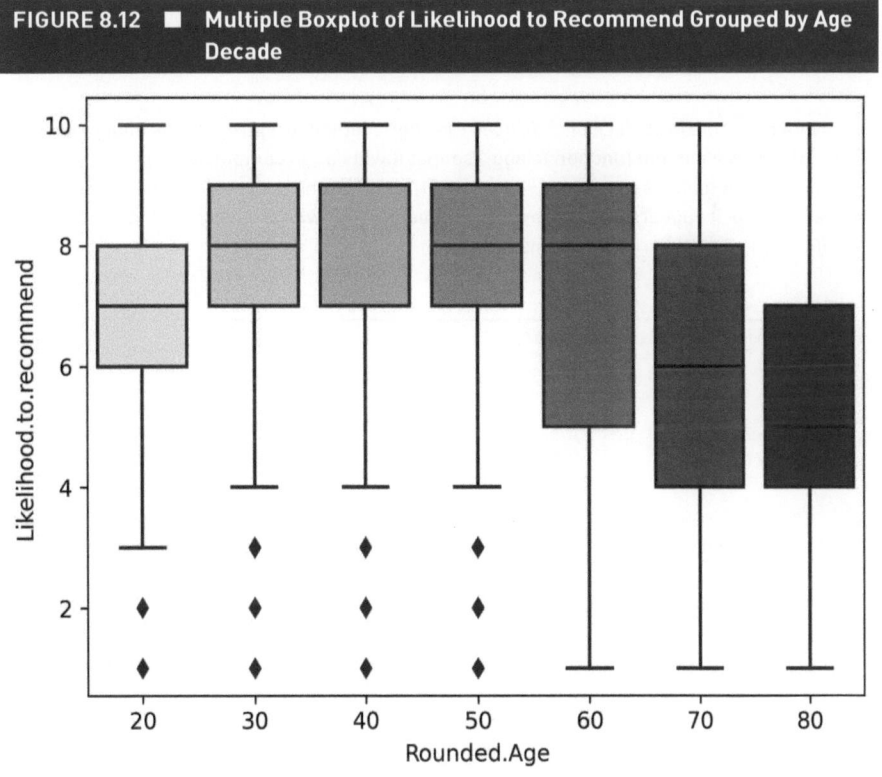

FIGURE 8.12 ■ Multiple Boxplot of Likelihood to Recommend Grouped by Age Decade

that we want with the help of the pandas groupby() method. This code uses Rounded.Age to group the pandas data:

```
survey_NPS = survey.groupby('Rounded.Age')\
             ['Likelihood.to.recommend'].agg([calc_NPS])

survey_NPS.reset_index(inplace=True)

survey_NPS

   Rounded.Age    calc_NPS
0       20       -12.412938
1       30        28.197027
2       40        34.083202
3       50        26.991590
4       60         2.868178
5       70       -40.302730
6       80       -53.803030
```

Notice how the first piece of code uses the agg() method to aggregate the data by groups. We can supply a custom function to agg() so that it will call to compute a summary value for each group. In this case we supply calc_NPS(), a function that we explored earlier in the book (a copy of it appears in this chapter's notebook). Now we can easily generate a bar chart of NPS values by age group with this code:

```
sns.barplot(x='Rounded.Age',
            y='calc_NPS', data=survey_NPS,
            color='tab:blue')
plt.savefig('8_13.pdf', bbox_inches='tight')
plt.show()
```

The resulting bar plot, which appears in Figure 8.13, paints a clear picture of age differences in NPS. People in their 20s as well as people in their 70s and 80s have negative NPS scores, indicating that overall, there are many more detractors than promoters. In contrast, the mid-range group, including people in their 30s, 40s, and 50s, have positive NPS scores, with people in their 40s higher than everyone else. We must take care in the interpretation of these data, however, because of previously observed differences in traveler groups. First, business travelers probably fall more commonly into those midrange groups, whereas younger and older people may predominate in personal travel for leisure and other purposes. Relatedly, business travelers in their 30s, 40s, and 50s may have both the experience and resources to tolerate travel delays better.

FIGURE 8.13 ■ Net Promoter Scores (NPS) for Various Age Groups

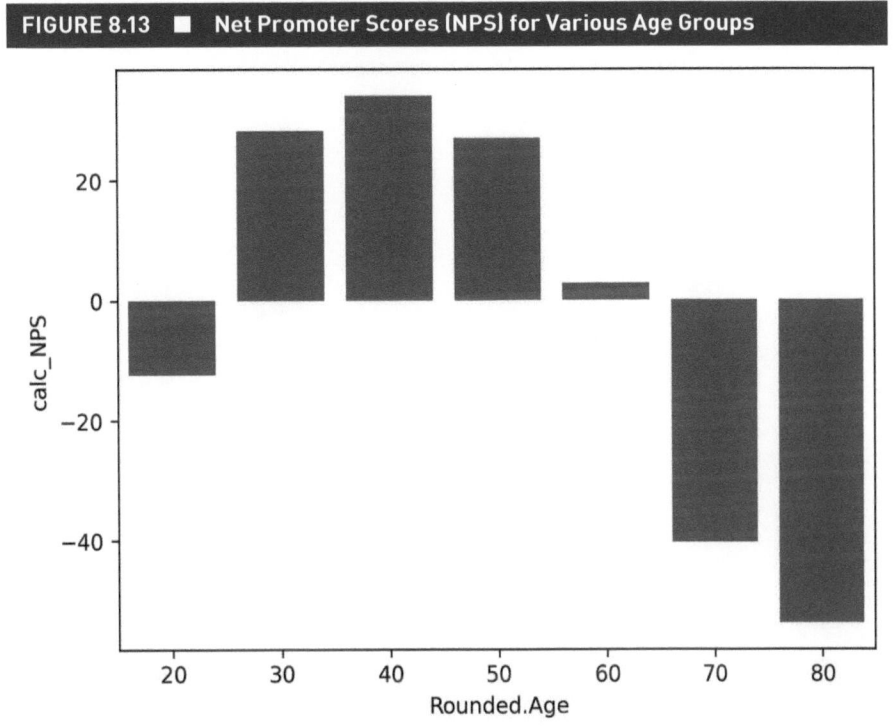

We can explore this latter point by adding another grouping variable into the mix. Here's a code cell that groups on both age decade and whether a delay was experienced:

```
survey_NPS_with_delays = \
        survey.groupby(['Rounded.Age', 'Big.Delay'])\
        ['Likelihood.to.recommend'].agg([calc_NPS])
survey_NPS_with_delays.reset_index(inplace=True)

survey_NPS_with_delays.head()

    Rounded.Age     Big.Delay          calc_NPS
0        20            False           2.227433
1        20            True          -39.732021
2        30            False          36.705362
3        30            True           12.433429
4        40            False          40.702494
```

This code cell provides a partial look at the resulting grouped DataFrame. Notice that there are two entries for each age group, one for the group who experienced delays and one

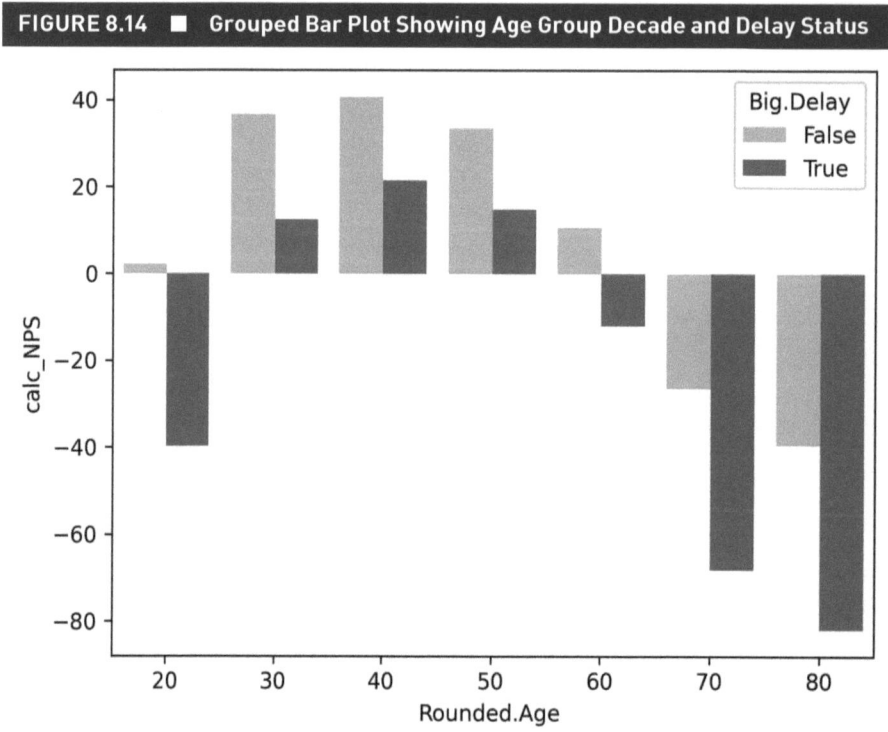

FIGURE 8.14 ■ Grouped Bar Plot Showing Age Group Decade and Delay Status

for the group who did not. This grouped DataFrame sets us up for a new version of the previous bar plot, which appears in Figure 8.14.

The code to produce the plot in Figure 8.14 appears below:

```
ax = sns.barplot(x='Rounded.Age', y='calc_NPS',
            data=survey_NPS_with_delays,
            hue='Big.Delay', palette="Blues")
plt.savefig('8_14.pdf', bbox_inches='tight')
plt.show()
```

The only difference between this code and the code that produced Figure 8.13 is that this code includes an argument to seaborn's barplot() function that says hue='Big.Delay'. This instructs seaborn to show two different bars for each group notated on the X-axis. In conjunction with the palette="Blues" argument, we get a light grey bar when Big.Delay is False and a dark grey bar when Big.Delay is True. The results are quite striking: Travelers in that middle group, with ages in the 30s, 40s, and 50s, show somewhat lower NPS scores, but as a group remain net promoters of the airline. Travelers in their 60s end up in opposite camps: Those who experience delays are net detractors, whereas those who do not experience delays are net promoters. Finally, the younger and older groups experience a massive increase in the number of detractors. When thinking about an interpretation of Figure 8.14, we might wonder whether the attitudes of personal and leisure travelers might be more susceptible to being adversely affected by flight delays.

One final graphic to close out this chapter will allow us to explore that question. In this case we will use a pandas capability known as a pivot (in the world of spreadsheets, sometimes called a "pivot table") to produce a heatmap of NPS across two grouping variables: age and travel type. Here's the code:

```
survey_NPS_with_name = \
     survey.groupby(['Rounded.Age', 'Type.of.Travel'])\
     ['Likelihood.to.recommend'].agg([calc_NPS])
survey_NPS_with_name.reset_index(inplace=True)

plt.figure(figsize=(14,7))

tmp_df = survey_NPS_with_name.pivot (index='Rounded.Age', \
                       columns='Type.of.Travel',\
                       values='calc_NPS')

sns.heatmap(data=tmp_df, annot=True, cmap="Blues")

plt.xlabel("Travel Type")

plt.savefig('8_15.pdf', bbox_inches='tight')

plt.show()
```

The first chunk of code does the same kind of aggregation by group that we did before, using calc_NPS() as our aggregation function. In this case, we are grouping the data by age decade as well as by travel type, which we know from before contains three groups. After a line of code to control the plot size of the figure, we then do a second operation: This code produces a temporary DataFrame using the pivot() method that pandas offers. To use the pivot() method, we must specify the index, which designates the variable that will control the rows of the resulting DataFrame, the columns variable, and the values variable that designates what value will appear in each cell. The resulting DataFrame is the perfect thing that we need to call seaborn's heatmap() function. Figure 8.15 displays the resulting heatmap.

You can control the color palette with the cmap argument to seaborn's heatmap() function. In Figure 8.15 we have asked for a palette with difference shades of blue, though they will appear grey in this book. This is not ideal—heatmaps usually look better with a palette that runs from red to another color such a green, where red shows the strongest or most adverse value of the underlying variable. In our case, Figure 8.15 shows a dark grey where the NPS has the greatest proportion of promoters and light grey or even white where we have the greatest proportion of detractors. You can tell what is happening here both by the specific numbers that appear in each cell (the NPS scores for each subgroup) and by the color bar legend to the right of the main diagram. This is one situation where the redundant information provided by the color scheme and the specific NPS values is welcome: The

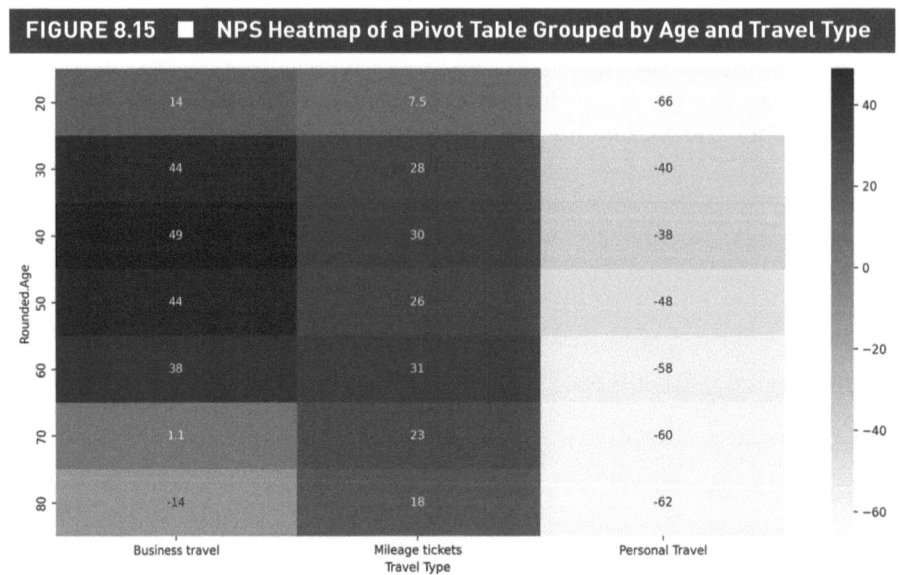

FIGURE 8.15 ■ NPS Heatmap of a Pivot Table Grouped by Age and Travel Type

color draws the eye to big-picture patterns, and the specific NPS numbers support a careful interpretation of contrasts among different cells.

Keeping the dark-light color scheme in mind, the eye is quickly drawn to the fact that business travelers in their middle years have the strongest NPS scores. In addition, it is clear that across the age spectrum, personal travelers are quite dissatisfied as a group, with the strongest levels of dissatisfaction among the youngest and oldest travelers who were surveyed.

It is important to note that the graphics we generated for this part of the case study were a valuable starting point in the exploration of this dataset but do not on their own represent a complete analysis. We are still at an exploratory phase in understanding this large dataset, and we would want to have histograms and other summaries of virtually every numeric variable in the dataset as well as tabular or grouped summaries based on other categorical variables. The sequence of graphics we generated for this chapter tells an interesting story about the age of passengers, the types of travel on which they were embarked, and their responses to delays, but there are almost certainly other interesting variables as yet unexplored.

CHAPTER CHALLENGES

(Problems 1–6 below require the use of the complete dataset. Refer to the code in this chapter's notebook for reading in the data from the JSON file.)

1. Write a paragraph that describes the last graph in this chapter (Figure 8.15). What could airline managers do about the data you described?

2. Create a bar chart of NPS by type of Type.of.Travel and Class. Which attribute do you think is more important in determining a person's LTR?

3. Create a bar chart of NPS value by Origin.State. Which state has the lowest NPS value?

4. Create a bar chart of NPS value by Destination.State. Which state has the lowest NPS value? In a sentence or two, compare these results to the results from the previous problem.

5. Create a scatterplot with Shopping.Amount.at.Airport on the X-axis and Eating. and.Drinking.at.Airport on the Y-axis. Describe in a sentence or two what the graph tells you about most people's spending behavior at the airport?

6. Create a new DataFrame that only contains rows where the person spent more than $400 on shopping at the airport. Redo the scatterplot from the previous problem, but this time set the color of the dot based on Likelihood.to.recommend. In a brief paragraph, describe what you see. Make sure to mention the number of rows.

7. Use the dataset from the previous question to create a barplot like the one in Figure 8.13 (with the bars representing different age groups). Instead of aggregating Likelihood.to.recommend with calc_NPS, use Eating.and.Drinking.at.Airport with the mean() method.

9 MAP MAGIC

LEARNING OBJECTIVES

Demonstrate the integration of disparate data sources and formats to produce a representative information model for decision-making.

Use the Folium library for generating map visualizations.

Plot geographic and numerical data within one visualization.

In many areas of human endeavor—environment, politics, education, business—we know that important events and activities happen in specific geographic locations. To understand these patterns of activity, organizations collect data that includes geocodes. At a minimum a geocode contains latitude and longitude information that together indicate a specific point on land or sea where the event or activity occurred. By examining these patterns of activity on a map, we may be able to draw helpful conclusions about the activities in question.

Thus, this chapter continues our exploration of data visualizations, this time focusing on examining geocoded data. One illustrative example that shows the value of geocoded data was HousingMaps (http://www.housingmaps.com), a web application that grabbed apartment rental listings from the classified advertising service Craigslist and plotted them on an interactive map that showed the location of each listing. HousingMaps worked so well that Craigslist eventually built its own map interface for finding rental properties.

Showing a map location where each rental property is located provides a much more convenient interface for users than a list of addresses: In the years following the introduction of HousingMaps, dozens of popular interactive applications such as Realtor.com, Hotels.com, and AirBnB.com have deployed powerful map-based interfaces within their services. Inside these companies, however, data scientists also use map-based data visualizations both to analyze business activity and communicate their analytic results to managers in an intuitive manner.

If you recall the visualization principles and rules of thumb that we discussed in Chapter 8, you will anticipate that we can do much more when displaying geocoded data than simply plotting a dot on a map. For example, when displaying rental property data, we could color-code the symbol depending upon whether an apartment was in a high-rise building, an apartment complex, or a multifamily home. We could have studio apartments be one shape and multi-bedroom properties be another shape. We could size the dot based on monthly rental price. As these examples suggest, map visualizations are conceptually similar to the visualizations we have previously examined but with the added geographical component of the displayed information.

Because we mentioned color-coding, this is a good moment to reiterate that when developing map-based visualizations, it is important to accommodate the many people who have some form of color blindness. One approach is to use a gray scale (a range from white to black). Another approach is to pick a color range that varies in intensity from light to dark (such as ranging from white to blue).

MAP VISUALIZATIONS BASICS

Before we get started with Python packages and code, we need to review a little geography—specifically the meaning of longitude and latitude. As shown on the right half of Figure 9.1, lines of latitude run horizontally across the globe. Latitude lines are parallel circles that go all the way around the planet, where 0 degrees is the equator; +90 degrees is the North Pole and 90 degrees is the South Pole.

As shown on the left side of the diagram, longitude lines are vertical arcs that connect the North Pole to the South Pole. Longitude lines are not parallel, and zero represents the prime meridian at Greenwich in the United Kingdom. The longitude line at +60 degrees lies eastward from Greenwich, and it splits Europe from Asia. The longitude line at −60 degrees lies

FIGURE 9.1 ■ Lines of Longitude (Left) and Latitude (Right)

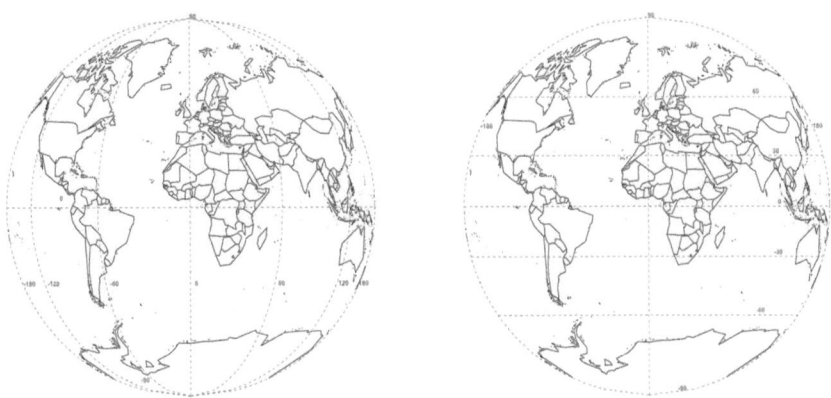

westward from Greenwich and passes through middle of South America. Interestingly, the longitude lines for +180 degrees and −180 degrees are exactly the same line. This line is in the middle of the Pacific Ocean. Longitude lines are only half circles, and that is why the overall range is +/−180: twice as many degrees as for latitude lines.

Take a moment to practice by choosing a few examples from the global map in Figure 9.1. For example, 0 degrees latitude (on the equator) and 0 degrees longitude falls right in the center of these maps and lies on a point in the Atlantic ocean underneath the west African country of Ghana. If you can see the area where you live somewhere on the map, try to estimate what the latitude and longitude are. After that, do a web search to find the specific longitude and latitude of your city or town.

There is another important idea to keep in mind when examining or creating maps. At small scales, for example, the area of a town or small city, longitude and latitude work like a regular Cartesian grid (a simple X and Y location on a rectangular field). At larger scales, however, the spherical shape of the Earth interferes with plotting the latitude and longitude on a flat surface. We use a technique called a "map projection" to render a spherical area onto a flat surface. All map projections create some kind of visual distortion because a spherical object does not want to be flat! Generally, however, we are used to seeing global maps with these distortions, so we don't normally think about it. For example, in the next two figures you can see two different map projections for rendering the whole globe on a rectangular grid. Figure 9.2 shows a Mercator projection: This projection was developed in the 1500s as an aid to ocean navigation. Using a map with this projection, ship captains can set one compass direction and reliably get to their destinations. The Mercator projection is extremely common and is the one you have most likely seen on global maps.

The next map shows a Gall-Peters projection. This one was developed in the 1800s in an effort to show the correct relative sizes of the earth's land masses. Note how in the Mercator

FIGURE 9.2 ■ Mercator Projection of the Surface of the Earth

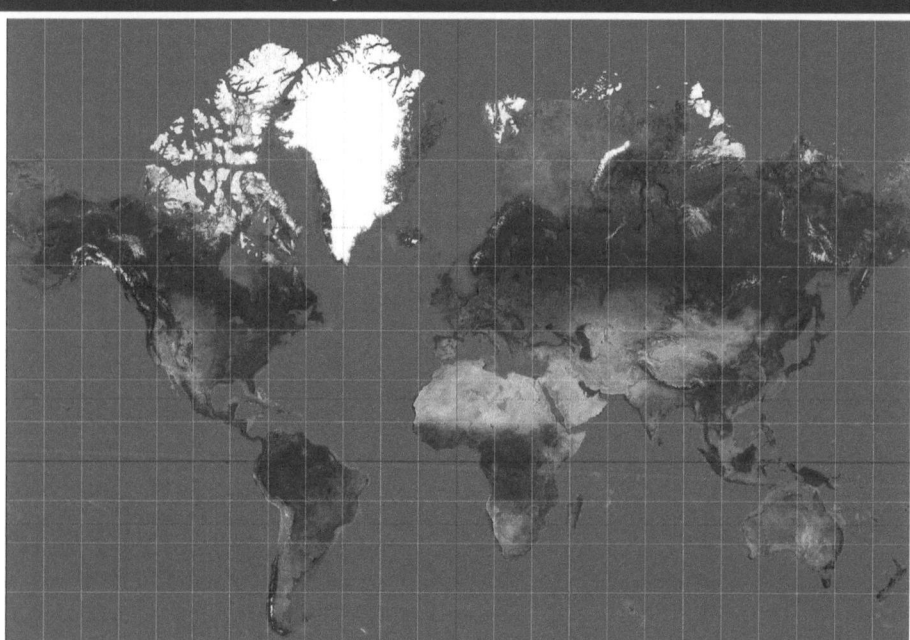

Note. "File: Mercator projection Square.JPG" by Strebe, https://commons.wikimedia.org/w/index.php?curid=17700069, licensed under CC BY-SA 3.0 DEED Attribution-ShareAlike 3.0 Unported https://creativecommons.org/licenses/by-sa/3.0/deed.en

projection, North America looks like it might be bigger than Africa. In the Gall-Peters projection shown in Figure 9.3, Africa correctly appears as being about 20 percent larger than North America.

You can notice in the Gall-Peters projection that the space between latitude lines gets more compressed as you go further north or south from the equator. This representation reflects the fact that the imaginary latitude line around the equator—the "belt" around the Earth—is the longest latitude line of all, and as we get closer to the poles, the circumference of the Earth at those latitudes becomes smaller. The key idea to remember is that whenever we render a map where the curvature of the Earth is large enough to matter, we need to choose a map projection based on the intended purpose of the map. As a rule of thumb, any map that represents a surface area larger than a small city may need to use a map projection.

CREATING MAP VISUALIZATIONS WITH FOLIUM

Now we are ready to start with generating a simple map using Python's Folium library. Note that there are variety of map libraries available for Python. We are using Folium because it provides an easy-to-use interface with a broad set of visualization options. Like many Python

FIGURE 9.3 ■ Global Map With Gall–Peters Projection

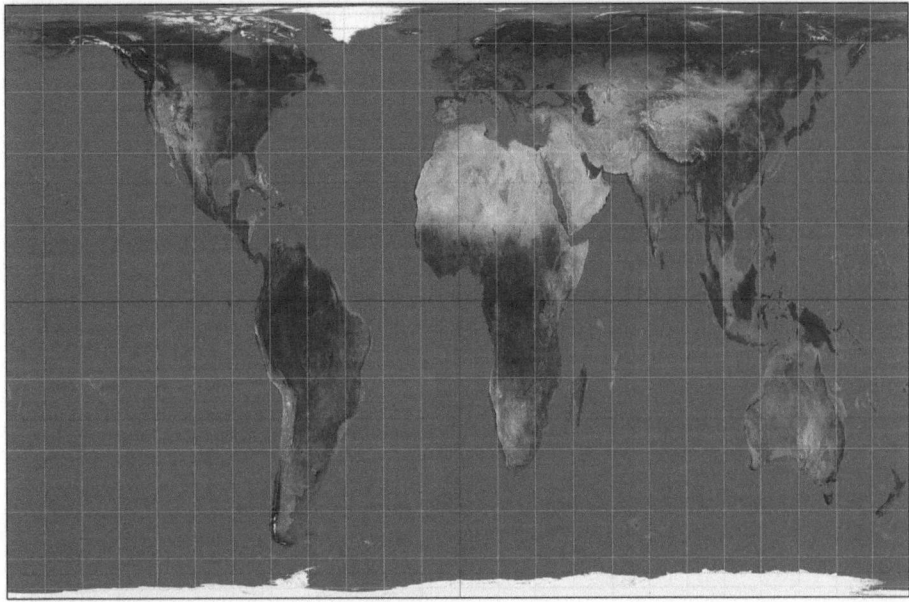

Note. "File: Gall–Peters projection SW.jpg" by Strebe, https://commons.wikimedia.org/wiki/File:Gall%E2%80%93Peters_projection_SW.jpg, licensed under CC BY-SA 3.0 DEED Attribution-ShareAlike 3.0 Unported https://creativecommons.org/licenses/by-sa/3.0/deed.en

packages, Folium is free and open source and was created and is maintained by a group of dedicated volunteers. As a data scientist, you may eventually work in situations where you need a level of customer support that cannot usually be provided for an open-source package, but for our purposes here, Folium is perfect.

Once Folium is imported, we can request a map using the folium.map() function. There are two arguments we will need to control this function. The first argument is the location, which is defined by specifying the center of the map, using latitude and longitude values. The second argument is how much the map should be "zoomed" by default. By specifying a location in the United States and a zoom value of 4, we can easily create a map that represents a substantial part of the North American land mass. We use the following lines of code:

```
m = folium.Map(location=[48, -102], zoom_start=4)
m.save("9_3.html")
m
```

Note that the first line of code calls folium.Map() to instantiate a map object and to store it in a variable called "m." The second line of code saves a copy of the map as an HTML file. You can download this HTML file to your own computer and examine the map in a browser.

If you looked under the hood at the HTML source code, you would find something interesting: There's no map in there! Instead, the HTML contains a copy of the latitude, longitude, and zoom level plus a bunch of calls to web services. Chief among these is "leaflet.js," a collection of open-source scripts for creating maps.

One implication of this is that your Folium map will work only if you are connected to the internet. Whether viewing the map within a Jupyter notebook or in your own web browser, the mapping scripts provided by leaflet.js have to be obtained from the internet before they can produce your map. If you are making a map for a presentation or report, grab a screenshot or use your browser to save a PDF copy of your map.

The last line of code has "m" on a line by itself, which causes the map to render in the notebook. Remember that by convention, the latitude number is usually offered first and the longitude second. Latitude 48 and Longitude −102 thus correspond to a North Dakota location that appears in Figure 9.4. By the way, a mnemonic to remember the difference between latitude is longitude is that "Lat is Flat," meaning that latitude lines are the horizontal markings on a typical map.

As you may have guessed, the map projection for this visualization was the Mercator projection. In fact, when using Folium, the default projection is Mercator. However, you can change the projection with the Coordinate Reference System (CRS), which is an optional parameter when creating a map. But for this chapter, we will continue to use Folium's default projection.

One of the advantages of using Jupyter notebooks is that they are rendered by a regular web browser—this means that we can take advantage of many of the interactive features that

FIGURE 9.4 ■ Part of North America, Centered Horizontally on a Location in North Dakota

Note. Rendered by the Folium package.

a browser offers and work on our Python code at the same time. For example, if you click and hold the mouse down on the map, you can drag the map (and see other parts of the world).

Although it was certainly easy to create, this map isn't particularly useful as a data visualization. A simple graphic like this is often called a "base map." Base maps become more useful when additional layers are plotted on top of the base. For our purposes, we want to add layers that represent geocoded data elements that we have available. For example, we can shade each state, based on some attribute, such as the population of the state. To do this, we can read in the state population dataset in the same way we have done in previous chapters. The code for this appears in this chapter's notebook.

In addition to the population data, we also need to give Folium the outlines of the different states as well as the names of the different states. With this information the Folium library can set the color for each state (based on the population data we provide) and the geometry of each state (based on the outline of the states that we need to provide). Luckily, there are JSON files that have this state shape information on the web. Here's some code that glues together the correct URL for accessing the shape data:

```
base_url = 'https://raw.githubusercontent.com/' + \
           'python-visualization/folium/master/examples/data'
state_geo = base_url + '/us-states.json'
state_geo

'https://raw.githubusercontent.com/python-visualization/folium/master/
examples/data/us-states.json'
```

We can pass this JSON shape data file to Folium, but before we do that, let's explore the shape file a bit, which shouldn't be too difficult because we already have some experience reading JSON files. There's code in the notebook for this chapter that reveals the contents of the entry for Alabama, the first state in the alphabetized list. The Alabama entry shows that the JSON file has a list of features, which includes points for a polygon (defining the shape of the state). This polygon data is the hidden superpower of vector-based maps: Instead of storing a picture of a state, county, city, or town, we store a series of points on the global map as latitude/longitude pairs. And like a connect-the-dots activity that you might have done when you were a kid, if we drew a line between each neighboring pair of points, we would have an outline of a geographic entity such as a state. Now if you live in a region that forms a nearly perfect rectangle—the U.S. state of Wyoming and the country of Egypt come to mind—you might not need much more than four points to represent the borders of the region. In contrast, most regions have irregularly shaped borders. Alabama, the example shown in the notebook for this chapter, has more than 30 latitude/longitude pairs in the list that defines the outline of the state.

Now that we have our population data and the shape data, we are almost ready to create our map. However, we have one last issue in that our population data has a column for state names but not state abbreviations. The state shape data is keyed by the two-letter postal code abbreviation. We need the population data from the U.S. census and the shape data to use the same key. There are several ways to do this, but we will add a column to the population DataFrame that contains the state abbreviations. To do this, we will use the "us" library (that stands for U.S. rather than the word us). Our Google Colab environment does not contain a preloaded version of the us library, so the notebook for this chapter includes a code cell that says, "! pip install us" on a line by itself. This calls the pip3 command included with most computers and asks it to install the us package so that our Python code can access its functions and classes. After that we can use the lookup() function provided by the us package to make a list of the state abbreviations we need:

```
import us

abbr=[] # Make an empty list to hold the abbreviations

for state in US_region_pops['state_name']:
    state_temp = us.states.lookup(state)
    if state_temp == None:
        if state == "District of Columbia":
            abbr.append("DC")
        else:
            print(state)
    else:
        abbr.append(state_temp.abbr)

if len(abbr) == len(US_region_pops):
    US_region_pops['state_abbr']= abbr
    print("Success!")

Success!
```

In this code we import the us package then create an empty list called "abbr." We are going to fill abbr with the state abbreviations in the same order as the list of states and territories stored in the 'state_name' column on our US_region_pops DataFrame. The for loop in the middle of this code cell takes care of that, but you will notice that there is error checking in there: If us.states.lookup() returns the value "None," that indicates that it was not able to find an abbreviation for the state we requested. We found, through trial and error, that us.states.lookup() does not recognize "District of Columbia" and so returns "None" when that request is made. The code checks this and adds "DC" to the abbr list when it is needed. To be extra defensive in our code, we also include in the last three lines a test to make sure that we have the same number of abbreviations as we do states. Only when we pass that test do we copy the contents of abbr into a new column on the DataFrame. As a confirmation, we print "Success" to the console when this step is complete.

Having done all this work, we are finally able to create a map with the fill color representing the population of state. We do this by telling Folium to fill each state based on the values found in the july11pop column of data. The base map is the same as the simple map we previously created. The technical name for this type of visualization is a choropleth map. The word "choropleth" comes from two Greek roots meaning "many regions." The usual definition is any map where regions are filled in with different colors according to a scheme that represents data about the region. This is similar to the idea of a heat map, where we use different colors—for example, ranging from red to green—to represent data such as temperature, altitude, rainfall, or any other numeric variable. Here's the code that creates our choropleth map representing the July 2011 population of the U.S. states:

```
m=folium.Map(location=[48,-102],zoom_start=4)

folium.Choropleth(geo_data=state_geo,\
                data=US_region_pops,\
                columns=['state_abbr','july11pop'],\
                key_on='feature.id',\
                fill_color='Blues',\
                fill_opacity=0.9,\
                line_opacity=0.1,\
                legend_name='Population'\
            ).add_to(m)

folium.LayerControl().add_to(m)
m.save("9_4.html")
m
```

Notice that the code begins with the same call to folium.Map() that we used earlier in the chapter to create the base map centered on North Dakota. The most important function call in this code cell is to folium.Choropleth(). Let's review the purpose of each of the parameters to the Choropleth() method:

- geo_data: the URL of the JSON file containing the shape data
- data: the DataFrame where the data values are stored
- columns: the two columns we need from the DataFrame, namely, the state two-letter postal code abbreviation (state_abbr; these are our key values for matching with the shape data) and the population data (july11pop)

- key_on: "feature.id," which tells the function how to locate the two-letter abbreviations in the JSON file that has the shape data; these abbreviations are the key data that allow matchups with the population data
- fill_color: the color scheme; we selected "Blues" to make a readable map in this two-color book; other available Python color schemes include "Greys" and "Reds"
- fill_opacity: level of transparency for the fill
- line_opacity: level of transparency for the outline of the states
- legend_name: title on the color key

Figure 9.5 shows the resulting map. In viewing the map, we can see the states with the highest population are California, Texas, and New York.

When seen in color, the labels on this map don't look quite right. It seems that when we plotted the shades of grey over the outlines of the states, part or all of the text labels for cities like Los Angeles and New York were underneath the choropleth layer. Because we chose a fill opacity that was slightly less than one, you can barely see the parts of the city names obscured by the fill color. Fortunately, Folium provides control over when and where labels get plotted.

SHOWING POINTS ON A MAP

Let's now add some points to the map. We can start by hard coding the latitude and longitude of a specific location. For example, we can create a made-up point somewhere in Texas and then add that point to the already existing map:

FIGURE 9.5 ■ Choropleth Map of the Continental U.S. Showing Population Values in Shades of Grey

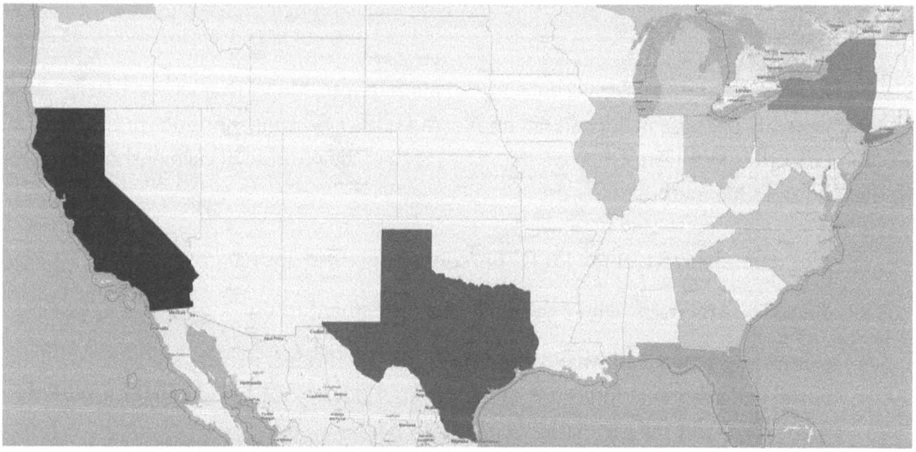

Note. Rendered by the Folium package

```
lat = 30
lon = -100
marker = folium.RegularPolygonMarker(\
         location=(lat,lon),
         radius=4,
         color='#000000',
         popup='my point in texas')
marker.add_to(m)
m.save("9_5.html")
m
```

It is important to realize that we are adding to our previous map using the code in this cell. After defining the latitude and longitude of a point, we use a function called "folium.RegularPolygonMarker()" to create a marker that we then add to the map using marker.add_to(). This function call specifies the latitude and longitude of our point, sets a size for the marker using the radius argument, sets the color to black using a hexadecimal code, and adds a pop-up help message. In Figure 9.6 we show a close-up of the map where a spot in southern Texas now contains a little dot.

The pop-up help message is not immediately rendered. You must click on the dot within the HTML-based display of the map for the point to appear. The shape, shading, and color of the point can all be controlled with additional arguments.

We can also show a point on the map using a named location. A named location is a specific address, city, or landmark that is converted into a latitude and longitude with the assistance of a geocoding service. This book is being written in Syracuse, New York, in the United States. Let's use geocoding to find where Syracuse, New York, is located on the map. To do this, we can use the geopy library, which makes it easy for Python developers to locate the map coordinates of named locations across the globe using third-party geocoders and other data sources. In our example, we use the Nominatim geocoding service, which is a free web service provided by OpenStreetMap (see https://nominatim.openstreetmap.org/ for more information on the service). OpenStreetMap is maintained by a cadre of volunteers and organized by a nonprofit foundation of that name that is based in the UK. Note that commercial geocoding services charge on a per-inquiry or subscription basis, and OpenStreetMap is one of the few available services that anyone connected to the internet can use for free. To keep the costs involved in providing geocoding manageable, Nominatim runs on donated servers. As such, we must limit the number of geocoding queries made to the service. As a rule of thumb, if an application will need to do more than one query per second, you should use a commercial geocoding service or set up your own Nominatim server. Here's the code:

FIGURE 9.6 ■ Choropleth Map of the Continental U.S. With a Close-Up on Texas Showing a Single Point With a Pop-Up Message

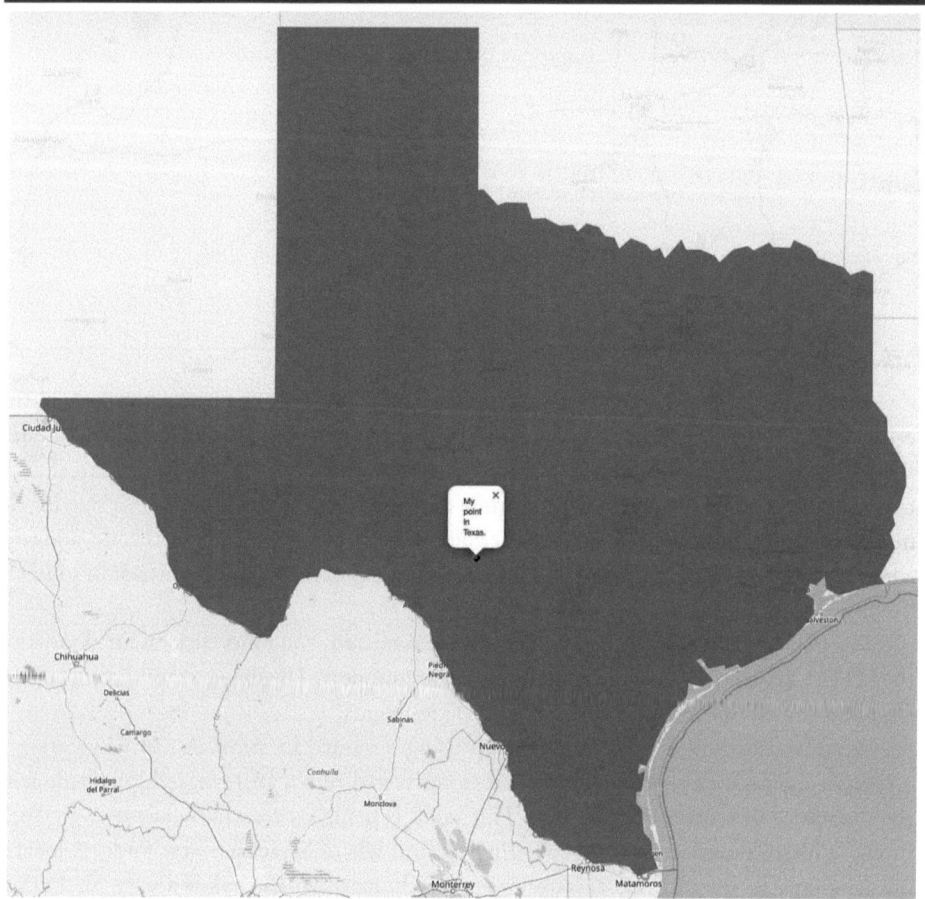

Note. Rendered by the Folium package

```
import geopy

locator=geopy.Nominatim(user_agent='IDSpython')

try:
   location=locator.geocode('Syracuse,NY')
   syracuse_location=[location.latitude,\
                      location.longitude]
except:
   syracuse_location=[43.088947,-76.154480]
```

```
circle_marker=folium.CircleMarker(\
        location=syracuse_location,
        radius=4,
        color='#000000',
        popup='Syracuse,NewYork')

circle_marker.add_to(m)
m.save("9_6.html")
m
```

The second line of code instantiates a locator class that can invoke the Nominatim service. Note that we have supplied a "user_agent" name that identifies this book. When you use the service, Nominatim recommends that you change the user_agent argument to contain your email address. The volunteers at Nominatim would contact you by email only if you exceeded the rate limits of the service.

The next five lines of code contain a "try:/except": Python construction that allows you to capture any errors that may occur when making the geocoding inquiry. Python first runs the code within the "try": clause. If any line of code in that section causes an error, Python switches to running the code in the "except": clause. This code represents the simplest use of the Python error management system, so there is much more flexibility available to handle more complex situations. Also note that we are continuing to use the choropleth map to which we previously added the dot in Texas. Our new map in Figure 9.7 contains the outlines of the states shaded to represent the population, the dot previously placed in Texas, and the new dot in Syracuse, NY.

Remember to check the terms of service for the geocoding service you use in your application to avoid exceeding rate limits. The U.S. Geological Service website has a searchable database for finding open sources of map data, and some of these sources can be used commercially.

You have likely noticed that the maps we have plotted so far look like the simple ones you get on your phone when you ask for driving directions. Everything we have done so far has been accomplished with "vector graphics," that is, with a set of line drawings filled in with colors. There is, however, a long tradition of using photographic data as the basis for a map, and such maps are sometimes referred to as "terrain" maps because they provide a detailed view of the land that the map depicts. As an alternative to vector graphics, "raster graphics" is a technique that uses millions of tiny dots to create a detailed image. When you use the maps on your phone, and you switch to what is sometimes called "satellite" view, what you are doing is replacing the vector-graphics base map with a raster-graphics base map that contains aerial photographs of the terrain. When viewing maps on a poor internet connection, you may also have noted that the terrain photographs appear gradually in a sequence of small squares. This is because raster graphics generally require much more data

FIGURE 9.7 ■ Detail of U.S. Choropleth Map Showing Circular Dot With Pop-Up Help Added to Syracuse, NY

than vector graphics. Each small, new square with a terrain image is filled in as it is received over your internet connection.

Folium provides a way to change the default background map to include terrain information—what are sometimes referred to as "stamen" terrain maps. Stamen maps are tiled images of varying levels of detail (the higher the zoom number, the more detail). For more information on stamen maps, you can check out the website of the San Francisco-based design firm that created them (http://maps.stamen.com). One last point is to notice, in the code, that to help make sure people can see the terrain, we changed the fill_opacity in the map to be 0.5 (rather than previously, where it was 0.9). This also helps make the labels more visible, as shown in Figure 9.8.

If you are using the notebook associated with this chapter, you may find it easier to examine the terrain features by using your mouse to zoom in on a particular section of the map. Figure 9.8 shows a zoomed-in view of the state of New York with sufficient detail that you can see the famous Finger Lakes south and west of Syracuse. The following code was used to generate this map:

FIGURE 9.8 ■ Detail of Choropleth Map of New York, Showing Terrain Base Map

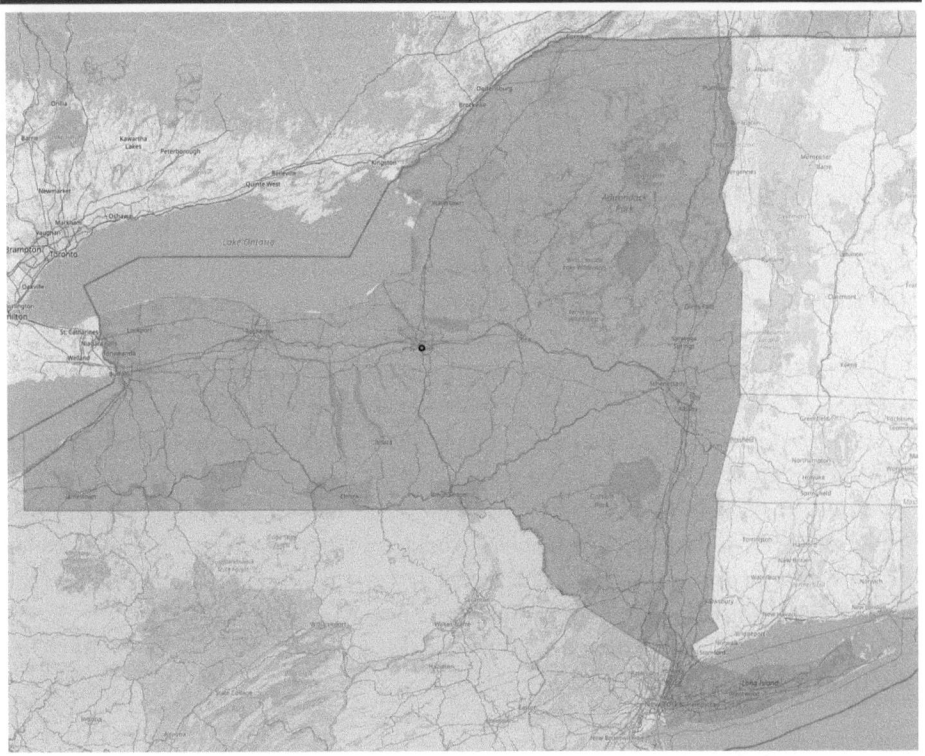

```
m = folium.Map(location=[48, -102], zoom_start=4,\
               tiles='Stamen Terrain')

folium.Choropleth(geo_data=state_geo,\
              data=US_region_pops,\
              columns=['state_abbr','july11pop'],\
              key_on='feature.id',\
              fill_color='Blues',\
              fill_opacity=0.5,\
              line_opacity=0.1,\
              legend_name='Population'\
            ).add_to(m)

folium.LayerControl().add_to(m)
```

```
marker.add_to(m)  # Add the point in Texas
circle_marker.add_to(m)  # Add the point in Syracuse

m.save("9_7.html")
m
```

For the first time in this chapter, we have created a new map: It was necessary to do this because the bottom layer of the map—the base map—is now a terrain rather than a vector map. We reused our marker and circle_marker objects to add the dots for Texas and Syracuse. The only important difference between this code and what we ran earlier was the addition of the "tiles='Stamen Terrain'" argument to the call to folium.Map(). This argument causes Folium to seek out the small, raster-graphics tiles from the internet and use those to fill in the base map. Note that as you use the zoom control on this map, a new set of tiles is fetched to represent the high level (for zooming in) or lower level (for zooming out) of detail needed to show the terrain.

The notebook code for this chapter also includes an additional example that shows how to plot the Citi Bike data onto a map of New York City. Given that both the bike availability data and the map data are freely available on the internet, this example demonstrates a way of quickly creating a web application that includes a mapping component. That final example completes our focus on visualizations. In the next chapter, we turn to linear modeling, our first effort to create predictive models.

CASE STUDY: EXPLORE NPS BY STATE AND CITY

```
Case Key Points:
 - Generate DataFrames that have NPS by state and by City
 - Visualize NPS values on a map
```

Airlines are among the many businesses whose product is intimately tied to questions of geography. Moving people and cargo from airport to airport involves a variety of geographically linked concerns such as the population of the surrounding area (which affects the number of flights), the popularity of areas such as tourist destinations, and issues of weather and terrain. For these reasons, representing customer satisfaction data on a map may reveal interesting insights. To start, we will again read in the large survey data we have previously used. A copy of the code for this appears in this chapter's notebook.

Next, we will create a DataFrame that has the NPS value for each origin city so that we can show NPS values on a map. There's nothing special about first choosing an origin city instead of a destination city: Both should eventually be analyzed. The code that follows uses the same techniques as we tested in previous chapters to group and aggregate our data:

```
grouped_by_city = survey.groupby('Origin.City')
NPS_by_city = grouped_by_city['Likelihood.to.recommend'] \
    .agg([calc_NPS])
NPS_by_city.reset_index(inplace=True)

NPS_by_city.head(2)

0       Aberdeen, SD            7.142857
1       Aguadilla, PR          -9.375000
```

We have summarized the data, generating (for each Origin.City) an NPS_by_city value. We used our CalcNPS() function that we first defined in Chapter 4. In addition, we need to have the longitude and latitude for each city. Each row of our survey already has that information, so a simple way to generate that data for the new DataFrame is to use the drop_duplicates() function to yield one entry for each city:

```
coordinates = survey[['Origin.City', 'olong',\
                      'olat']].drop_duplicates()
print (coordinates.shape)
coordinates.head(2)

(212, 3)

         Origin.City        olong          olat
0        Minneapolis, MN    -93.27980      44.97816
20       NewYork, NY        -73.97022      40.71836
```

The output shows that we have 212 cities in total. Each row of the resulting DataFrame has three columns, one that names the city and two for the latitude and longitude. Two example cities are shown. One additional step combines the DataFrames using the merge function. The key field that connects the two DataFrames is the origin city, and we should end up with one additional column representing the NPS for each city:

```
city_NPS = pd.merge(NPS_by_city, coordinates, on='Origin.City')
print (city_NPS.shape)
city_NPS.head(2)

(212, 4)

        Origin.City    calc_NPS      olong        olat
0       Aberdeen, SD   7.142857     -98.47103    45.46319
1       Aguadilla, PR -9.375000     -67.04100    18.37071
```

This output confirms that we now have four columns and provides two example cities. When using data like this in a commercial application, we would want to use these as test cases to verify that the coordinates and the NPS calculations are correct. The notebook for this chapter contains additional code that aggregates NPS values by state so that we can create and fill a choropleth map. We also reuse some lines of code from earlier in the chapter to look up two-letter state abbreviations needed to match with vector-based outline data for each state. Now that we have all this NPS data, let's first create the choropleth map.

```
m_choropleth = folium.Map(location=[48,-102],zoom_start=4)

myscale = [i for i in range(-100,100, 20)]

folium.Choropleth(geo_data=state_geo_url =
            'https://raw.githubusercontent.com' + \
            '/python-visualization/folium' + \
               '/master/examples/data' + \
            '/us-states.json'
         print(state_geo_url), \
               name='choropleth', \
               data=state_NPS, \
               columns=['state_abbr', 'calc_NPS'], \
               key_on='feature.id', \
               fill_color='Blues', \
               fill_opacity=0.7, \
               line_opacity=0.2, \
               threshold_scale=myscale, \
               legend_name='NPS by state' \
         ).add_to(m_choropleth)

m_choropleth.save("9_8.html")
m_choropleth
```

This code has a few minor, but interesting, changes over what we accomplished earlier in the chapter. Notice that the second line of code creates a new variable, myscale, that contains a set of values starting at –100 and proceeding upward in steps of 20. This defines that theoretical range of NPS from 100 percent detractors to 100 percent promoters. The myscale variable is used in the threshold argument to the folium.Chropleth() function call. Folium will use this list of numbers to create the color map and will also provide a legend with a bar that shows the mapping of colors to NPS values. There's also an argument that provides a label for the color map bar. We did this so that we could have a true sense of NPS values relative to the theoretical range values. If we did not define our threshold scale, then Folium would automatically scale from the minimum to the maximum that it found in the data. Additionally, we have reduced both the fill opacity and the line opacity so that labels will show through better. The resulting map appears in Figure 9.9.

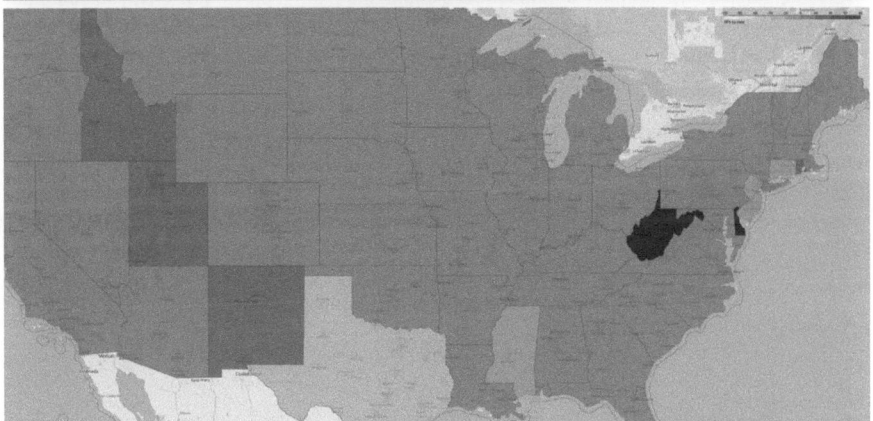

FIGURE 9.9 ■ Choropleth Map of the United States, Depicting State by State NPS Values

Note that Figure 9.8 shows that there is no NPS data for either West Virginia or Delaware, so these are filled with black on the map. We can also show the NPS value for cities on top of this choropleth map. There are three parts to the code. First, we must create a scaled version of our city-by-city NPS values because these values will control the radius of the marker we plot. We loop through all cities, and for each city, create a circleMarker and then add it to the base map:

```
max_nps=max(city_NPS['calc_NPS'])

city_NPS['scaled_nps']=\
    city_NPS['calc_NPS']/max_nps*10

for index, row in city_NPS.iterrows():
    folium.CircleMarker(location=(row["olat"],
                                  row["olong"]),
                        radius=row['scaled_nps'],
```

```
                        fill=True,
                        fill_color='white',
                        fill_opacity=0.8,
                        color='black',
                        weight=1,
                        popup=row['Origin.City']
                        ).add_to(m_choropleth)

m_choropleth.save("9_9.html")
m_choropleth
```

This code does not create a new Folium map, so all of the markers will be added on top of the existing choropleth map that we created in the previous step. The radius of the circle is controlled by a scaled value of the NPS for each city. If you wanted to depict an additional variable, such as the number of passengers departing from a particular city, you could use that variable to control the fill color of the circle. In this case, we fill each circle with white, using a black outline, to make it suitable for this two-color book. The result appears in Figure 9.10.

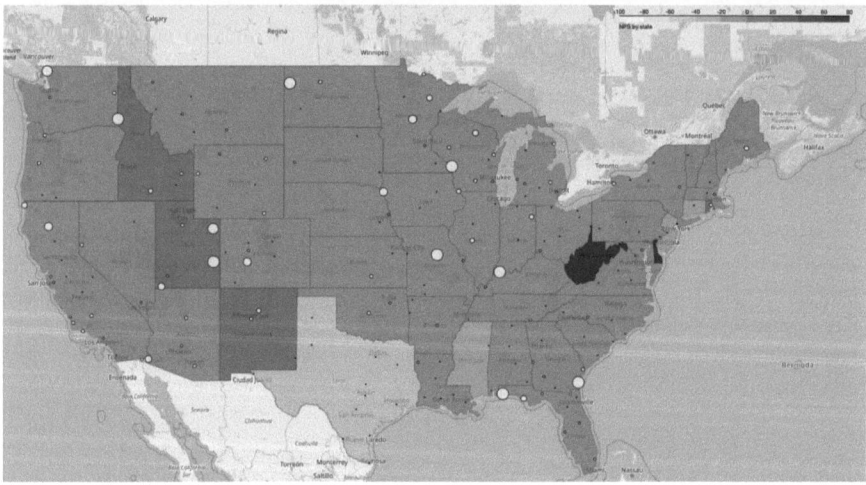

FIGURE 9.10 ■ Choropleth Map of the United States, Depicting State-by-State NPS Values, With Circles Representing the NPS for Origin Cities

There are a variety of ideas that can be inferred from the combined map in Figure 9.9. The three western states exhibiting the highest NPS values—Idaho, Utah, and New Mexico—all benefit from the fact that they have only a few departure cities and that NPS values are reasonably high for those cities. Two southern states, Texas and Mississippi, have low NPS values for all of their departure airports and therefore low NPS values overall. Of the 212 departure cities across the United States, there are only about a dozen that have quite high NPS values, and only a few of these are in large population centers. These positive examples should be investigated to see if the data can yield insights into possible reasons for high passenger satisfaction. In contrast, the states down the East Coast with high population densities, from Massachusetts down to Virginia, lack any high NPS departure cities. Taken as a whole, this map provides a high-level view of airline passenger satisfaction across the United States. The map can be used to direct attention to trouble spots and positive exemplars for further exploration and analysis.

CHAPTER CHALLENGES

1. Go to syracuse.craiglist.org, and click on the apts/housing for rent link. Switch to the map view and search for "Syracuse University." Review the map, and report how many listings there are near the university.

2. Explore https://github.com/python-visualization/folium/tree/master/examples/data, and report a list of maps that are available on that location.

3. Find the longitude and latitude of the center of New Zealand (using geopy).

4. Display a map of New Zealand using the location identified via geopy.

5. Generate a map showing the number of available bike slots in New York City.

6. Using the case study data and code, generate a choropleth map of NPS values for the *destination* states.

7. Find a census dataset with additional information (https://www2.census.gov/programs-surveys/popest/datasets/2010-2016/national/totals/nst-est2016-alldata.csv). Parse the dataset, and then display the data in a useful manner using the different visualization techniques covered in these last two chapters.

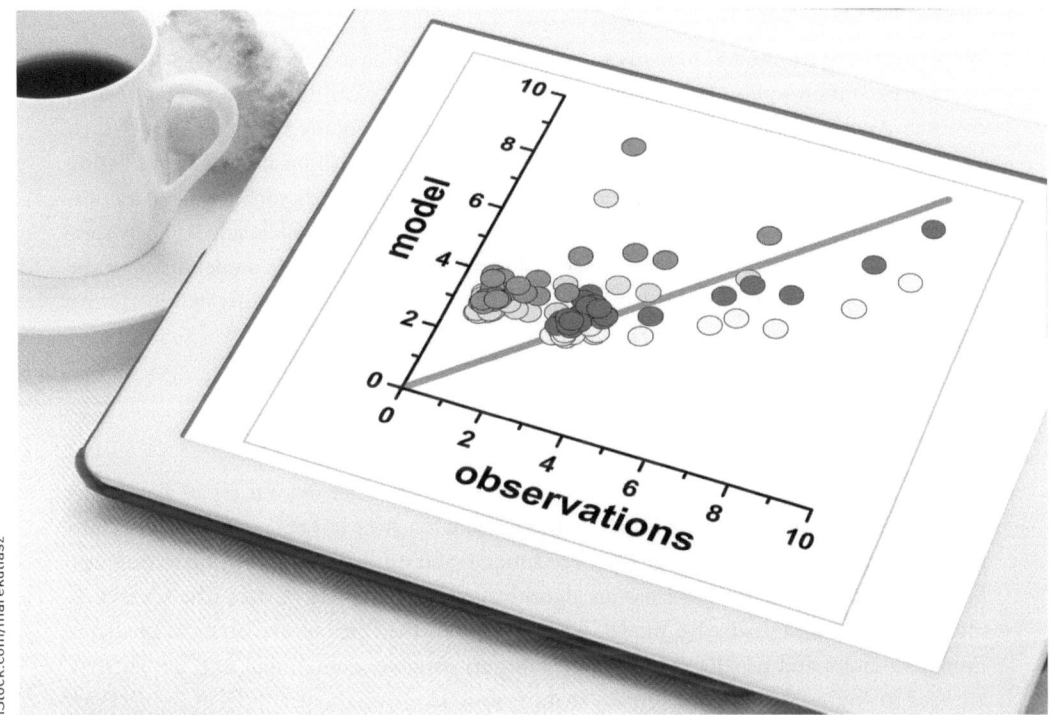

10 LINEAR MODELS

LEARNING OBJECTIVES

Understand the difference between supervised and unsupervised learning.

Understand the difference between regression and classification.

Explain and be able to use cross validation techniques.

Build and use models to interpret and understand the data.

Utilize the sklearn package to create and interpret linear models.

Use the Caret package to build linear models.

Interpret R-squared as a metric of model quality.

WHAT IS A MODEL?

People use the word "model" in many ways. Some have nothing to do with data, such as using the term to connote a person who might appear in a magazine wearing a certain sweater, shirt, or watch. Another use of the word describes a small-scale representation of larger objects (such as a model railroad). In this chapter, we won't discuss either fashion models or miniature trains. Instead, we will use data to explore the connection of a predictor variable to an outcome variable using some simple math. We will discuss predictive models calculated by software such as that provided by Python's SciKit-Learn package. Predictive models uncover the relationship between an input and an output variable in a way that supports making future predictions using new observations of the input variable.

SUPERVISED AND UNSUPERVISED LEARNING

"Statistical learning" and "machine learning" are terms that refer to a variety of algorithms that process data and find patterns in it. Statistical and machine learning techniques fall into two broad categories: supervised learning techniques and unsupervised learning techniques. Supervised learning is when we use an algorithm on an initial set of data (the supervised phase) to establish a predictive model. The data are in the form of one or more predictor (input) variables and usually one outcome (output) variable. Some input and output data are used to "train" the model, a process that involves adjusting the strengths of connections among variables to maximize the model's ability to predict the output from the inputs. We then use the resulting trained model to make predictions on new data. The model is thus a recipe for taking in some data and predicting some results. The connections between variables are referred to as coefficients.

Unsupervised learning is when we use an algorithm to group, organize, cluster, or otherwise understand patterns among a set of variables. Unlike supervised learning, there is no outcome variable to predict. Instead, an algorithm uses one of a variety of methods to combine or organize the input variables into a more compact and interpretable representation. In short, the unsupervised algorithm uncovers patterns on its own with no prior knowledge of what the categories should be or how the data could be reorganized.

An important supervised learning technique is called "regression." The simplest form of regression is "linear regression," so called because the key connection between an input variable and an output variable is a slope that describes the position of the best-fitting line that makes that connection. A type of math called "matrix algebra" allows linear regression to transform a dataset into a linear equation that specifies the slope for each predictor variable plus one additional coefficient known as the intercept. These linear equations allow us to make predictions for numeric outcome variables.

Let's make this thinking more concrete. As an example, we could use the number of oil changes for a car to predict the repair costs for that car. All we would need to start the analysis is some previous data on oil changes and repair costs. Linear regression will calculate a slope and intercept to go with the number of oil changes. We can think of the result graphically as the position of a line that slices through the middle of a group of points. Once we have the computed value of the slope and intercept, we can use them to predict future repair costs based on the number of oil changes we decide to do on our car. Linear regression works well in this situation because the number of oil changes is an integer and the repair costs are represented numerically as dollars or some other currency. Both variables are ordered, numeric measurements of a particular quantity.

Another important prediction technique is called "classification." The only notable difference between regression and classification is that the outcome variable in classification is categorical. Some examples of categorical outcomes are 1) whether you get hired for a particular job; 2) whether you attend a concert; 3) whether you drive a sedan, SUV, or minivan; or 4) the continent on which you were born. In each of these examples, a piece of data would be recorded either as a brief descriptive string (e.g., "hired" or "not hired") or by using a number to stand in for the name of a category (e.g., hired=1, not hired=0). The first two examples are binary classifications because there are only two possible outcomes. The second pair of examples are known as "multi-class" because there are more than two categories.

In this chapter, we will examine linear regression, which as previously noted, can be considered as a simple supervised regression learning technique. In the next chapter, we will examine supervised learning for classification. Two chapters ahead from here, we will explore unsupervised learning.

LINEAR MODELING

Finding relationships between variables is one of the basic aims of data science. The question "Does X influence Y?" is of prime concern for data analysts from every sector: Are house prices influenced by incomes? Is the growth rate of crops improved by fertilizer? Do taller sprinters run faster? We can answer these kinds of questions by building predictive models. One of the simplest forms of predictive model is called a linear model because it uses the slope and the intercept of a line to represent the relationship between an X (predictor) variable and a Y (outcome) variable. This method is often used by statisticians and is known as linear modeling. The term "linear model" actually covers a wide variety of methods from the relatively simple to the sophisticated, but all these methods create models that can be used for prediction. You can get an idea of how many different methods there are by looking at the regression analysis page in Wikipedia and checking out the number of models listed on the right-hand sidebar (and, by the way, that list is not exhaustive).

The original ideas behind linear regression were developed by some of the usual statistical suspects discussed in Chapter 5, such as Gauss, Galton, and Pearson. The biggest individual contribution was probably by Gauss, who used the procedure to predict movements of other planets in the solar system when they were hidden from view and to correctly predict when and where they would appear again.

The basis of all these methods is the idea that it is possible to fit a line to a set of data points. That line then represents the connection between an independent variable (aka predictor) and a dependent variable (aka outcome). It is easy to visualize how this works with one variable changing in step with another variable. The chart in Figure 10.1 shows a line fitted to

FIGURE 10.1 ■ The Best-Fitting Regression Line Intersecting a Set of Points

Note. From *An Introduction to Data Science*, by J. S. Saltz and J. M. Stanton (2017), p. 198. Copyright 2018 by Sage Publications. Reproduced with permission.

a series of points using the so-called ordinary least squares method (a relatively simple method of finding a best-fitting line). The chart shows how the relationship between an input (independent) variable on the horizontal X-axis—relates to the output (dependent) values on the Y-axis. In other words, the output variable is dependent (is a function of) the independent variable.

Note that although the line fits the points fairly well, with some points above and some points below the line, none of the points are precisely on the line—the training data do not fit the line precisely. As we discuss the concepts in regression analysis further, we will see that understanding these discrepancies is as important as understanding the line itself. The mathematical idea that allows us to create lines of best fit for a set of data points like this is that we can find a line that will minimize the sum of the squared distances between the line and each of the points. Once the model calculates the line that minimizes the some of the squared distances for all the points, the model represents the line with an equation you might remember from algebra:

$$Y = bX + a$$

Do not worry if you have never taken an algebra course: All you need to understand is that the equation has two values that describe the line (b and a). The letter b denotes the slope of the line and the letter a describes where the line crosses the Y-axis, also known as the "Y-intercept." Sometimes people refer to these values generically as "coefficients." These values are calculated by means of the matrix math behind linear regression as applied to the training data. Note that a linear model requires a minimum of two observations (in the simplest case, two X, Y pairs). In practice, most datasets used for training a linear model have dozens, hundreds, or thousands of observations. Once b and a have been calculated, the equation can then be used to predict a new Y value that might result from any given X value that you supply.

Although the mathematics behind these techniques can be managed by anyone with a little knowledge of matrix algebra, the reality is that with even with only a few data points, the process of fitting a line with manual calculations becomes tedious quickly. This is why software programs and tools, including Python, all include capabilities to conduct linear modeling. Because these tools are so widely available and so easy to use, we will not discuss the specifics of how the calculations are done but will move quickly to an example of how we use Python to do them.

To generate a linear model in Python, we can use the linear_model module from the Sci-Kit Learn (sklearn) package. Other popular Python packages for regression include NumPy and statsmodels. To create a linear model, we first create an empty model by instantiating the LinearModel class and then use the fit function provided by the class to create a model based on our training data. The fit() function, when given a list of points (X and Y values), generates the best-fitting line for the data using the ordinary least squares methods. In other words, fit() will calculate the appropriate b and a based on the training data. After that we can use the predict() function: When the model is given a new X value, it will use the equation of a line (Y = bX+a) to calculate a predicted Y value.

AN EXAMPLE—CAR MAINTENANCE

Is changing the oil of a car more frequently a good thing? Does it save money in the long run? Let's use linear regression to address this question. Let's say that we were put in charge of maintaining a fleet of company cars. We know that the company replaces the cars every 3 years, and in fact the company just replaced the cars (i.e., bought a new fleet of cars). The person who was previously in charge of the car maintenance didn't have a schedule for when to change the oil but rather changed the oil whenever the car happened to be available. Luckily, good maintenance records were kept, and now we want to try and figure out, using a data-driven approach, if changing the oil frequently is a good idea. The following Python code defines our dataset:

```
oil_changes= [3,5,2,3,1,4,6,4,3,2,0,10,7,8]
repairs= [300,300,500,400,700,420,100,290,
         475,620,600,0,200,50]

miles  = [20100,23200,19200,22100,18400,23400,
         17900,19900,20100,24100,18200,19600,
         20800,19700]
```

```
oil = pd.DataFrame({'oil_changes': oil_changes,
                    'repairs': repairs, 'miles': miles})
oil.head()
```

	oil_changes	repairs	miles
0	3	300	20100
1	5	300	23200
2	2	500	19200
3	3	400	22100
4	1	700	18400

Let's peek at the data for these cars: We have three columns of information. First, there is the number of oil changes that each car had during the past 3 years. Next we have the total amount of repairs (in dollars). Finally, we have the miles driven by each car.

Given these data, can we determine if changing the oil frequently is a good thing? Our independent variables are oil_changes and miles. The dependent variable (the one we are trying to predict) is repair costs. Before building a model, we should do some exploratory analysis, which professional data scientists always do prior to building a model. In this case, we can display (plot) the points with the following Python code:

```
plt.scatter(oil_changes, repairs)
plt.xlabel('oil changes')
plt.ylabel('repairs')
plt.show()
```

In this code, we use the plt.scatter() function from matplotlib to create a scatterplot. A scatterplot, also sometimes called a "scattergram," is simply a plot depicting a set of points on a two- or three-dimensional grid. Most of the time we stick with two dimensions for simplicity, as shown in Figure 10.2.

We can see a pretty clear pattern: The repairs are high when the number of oil changes is low, and the repairs are low when the number of oil changes is high. In the notebook for this chapter, a scatterplot also appears for the other predictor—miles driven—in relation to repair costs. For that variable, we don't see as clear a pattern as we did with oil changes. So to start simply, let's build our first model with only repairs as the outcome variable and oil_changes as the predictor.

```
from sklearn import linear_model
import numpy as np

X = oil['oil_changes'] # Put X into a 2D array
X = np.array(X)
```

```
X = X.reshape(-1, 1)

y = oil['repairs']

lm = linear_model.LinearRegression()

lm.fit(X, y)

LinearRegression
LinearRegression()
```

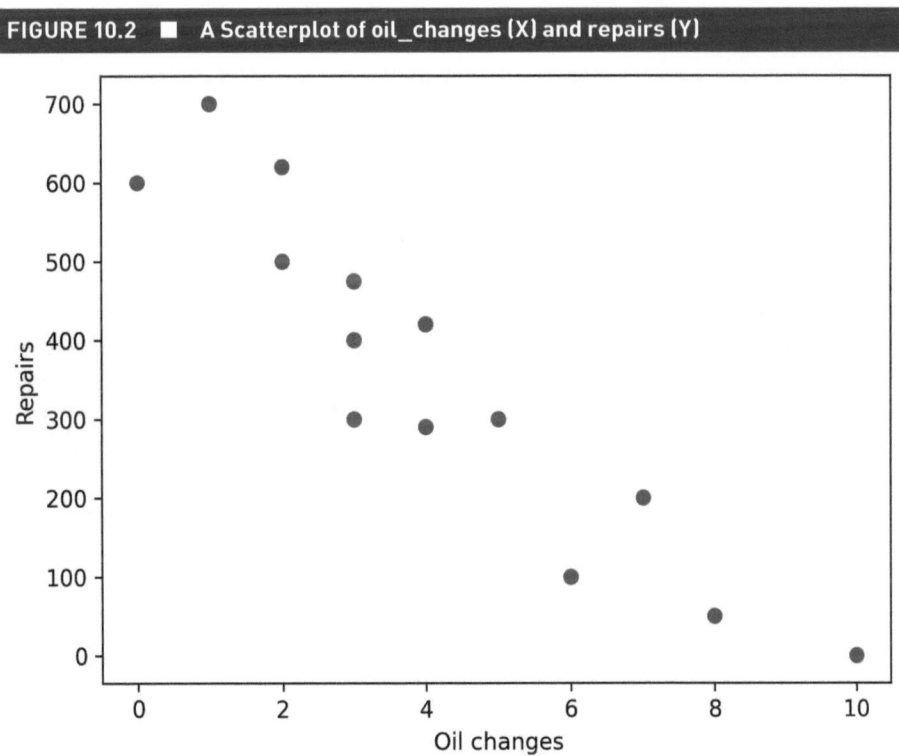

FIGURE 10.2 ■ A Scatterplot of oil_changes (X) and repairs (Y)

Note that in this code, even though we are creating a model with only one predictor (oil_changes), we need to pass that list of values as a two-dimensional NumPy array (a 2-D matrix) rather than a simple list of numbers. The reason for this is that the linear modeling function generally expects to use multiple columns of X values together as a set of predictors. To accommodate these expectations, lines three through five convert the series into a NumPy array and then use the reshape command to convert it to a 2-D array. The parameters for the reshape are the number of rows and columns. In this case, we know we want one column and then let the reshape() method decide how many rows, via the '-1 parameter. By the way, there's a convention among programmers to use a capital letter when a data object is a numeric matrix and a small letter when it is a "vector" (i.e., a simple list) of numbers. We have used that convention here by calling our predictor matrix X and our outcome variable y. The names, of course, make no difference to the lm.fit() function at the end of the code cell. The output simply confirms that the call to lm.fit() produced a regression data structure for future use. This data structure contains the slope and the intercept that we will need.

Next, we print the two key values that define our finished regression model:

```
print('Coefficients (slope): \n', lm.coef_)
print('Intercept: \n', lm.intercept_)

Coefficients (slope):
 [-71.99438202]
Intercept:
 652.1910112359551
```

The intercept and slope terms of the linear model are shown in the output (remember the equation Y = bX+a: the intercept is a, and the slope is b). The coefficient is the slope, the influence that oil changes have on vehicle repairs. The slope in our model has a negative sign, so it is saying that for every additional oil change, the expected repairs go down by $71.99. It might help if you ask yourself, "A one unit change in X leads to how much change in Y?" In this case an increase in X of 1 oil change leads to a decrease in Y of $71.99 in repair costs. The intercept, in our example, is essentially the expected value of the repairs when there were zero oil changes. This value suggests that if the company never changes the oil for a car, repairs will be on average about $652. One last thing about this output: The intercept shows as a single value, but the slope is inside a one-item list. The coefficients are provided as a list because it is often the case that we will have more than one predictor and therefore more than one slope coefficient.

We can also use the model to predict new values of Y and then see how well the model performed.

```
from sklearn import metrics

# Make predictions
y_pred = lm.predict(X)

# R-Squared -- the explained variance
r2 = metrics.r2_score(y, y_pred)
print('R-Squared:', round(r2,2))
```

```
# Root mean squared error
mse = metrics.mean_squared_error(y, y_pred)
print('RMSE:', round(np.sqrt(mse),2))

R-Squared: 0.87
RMSE: 76.58
```

In the call to lm.predict(), we are using the predict() method on our trained model, reusing all of the predictor data to generate a set of predicted values for y. We can then compare the predicted Y values to the actual Y values to decide whether we are satisfied with the quality of this model. We have two values that measure model quality: The first is called the "R-squared value," also known as the "coefficient of determination." The R-squared value can be interpreted as the proportion of the variation in the dependent variable that is accounted for by the complete set of independent variables (in this case, one independent variable). An R-squared value of 1.0 would mean that the X variable(s) made perfect predictions of the Y variable—every point in the training data fell right on the line. An R-squared value of zero would indicate that the X variable(s) did not predict the Y variable at all. R-squared cannot be negative. In this output, the R-squared is reported as 0.86, which means that the oil_changes variable accounts for about 86 percent of the variation in repair costs. Note that there is no absolute rule for what makes an R-squared good. Much depends on the purpose of the analysis and the nature of the data. In the analysis of human behavior, which is notoriously unpredictable, an R-squared of 0.10, 0.20, or 0.30 might be considered quite good. In this case, because we are working with a mechanical device whose operation depends upon materials engineering and physics, the adjusted R-squared of 0.86 suggests a solid result that could probably provide some actionable insights for the fleet manager.

Although the R-squared is perhaps the most important output to evaluate from a regression model, the second output helps us understand the level of accuracy we can expect to get. The root mean squared error indicates the average difference—above or below—between a predicted Y value and an actual Y value from the data. As such it is a general indicator of how inaccurate our model is likely to be on future predictions, assuming that any new data we get follows the patterns exhibited by old data. This latter point is particularly important for data scientists. Conditions can change, and when they do, old predictive models might not work anymore. For example, if the next set of cars purchased for the fleet are plug-in hybrids, our previous understanding of how many oil changes might be needed could go right out the window. In the present example, the root mean square error is $76.58, so we should not be surprised to find that future predictions of repair costs will typically be off base (too high or too low) by about this much. Root mean squared error is always calibrated on the same scale as the Y variable.

Next, we can show the line of best fit (based on the model) as an overlay on the scatterplot. This kind of graph helps with our understanding of the model results:

```
oil['y_pred']   = y_pred

oil_sorted = oil.sort_values(by=['oil_changes'])

y = oil_sorted['repairs']
x = oil_sorted['oil_changes']

plt.scatter(x, y)
plt.plot(x, oil_sorted['y_pred'], color='black')
plt.show()
```

The first line of code creates a new column on the original DataFrame to store the predicted values. This is not strictly necessary but might come in handy later. Likewise, the second line of code, which sorts the rows based on oil changes, is in case we might want to examine the data in a table: The sorting has no effect on the plot. We use plt.scatter() to show the original data as dots on the X-Y grid, but in a second step we overlay a simple line representing the predicted values using plt.plot(). The resulting graphic appears in Figure 10.3.

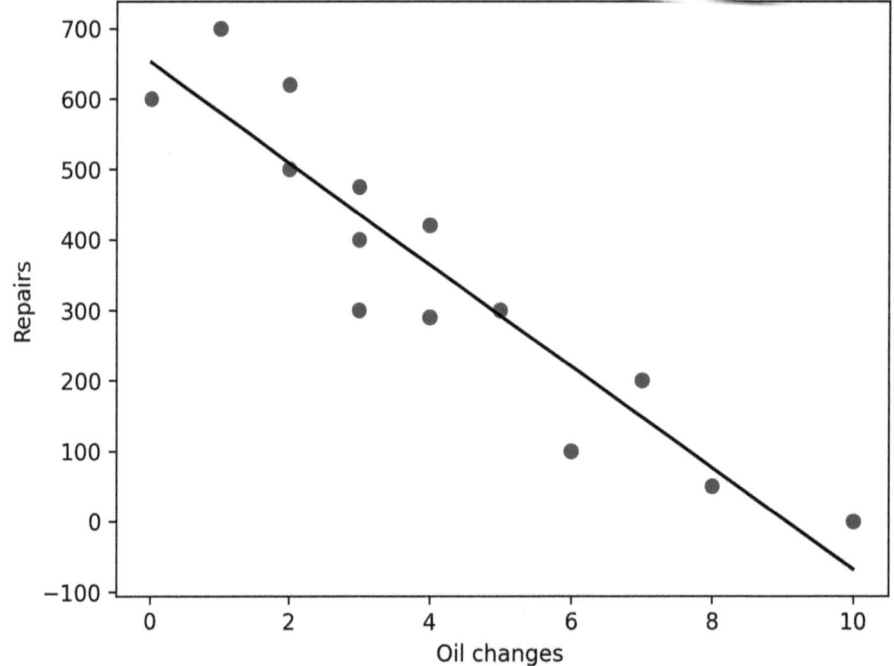

FIGURE 10.3 ■ Scatterplot of Oil Changes Versus Repairs With Line of Best Fit

This model suggests that we should do as many oil changes as possible. For example, the line predicts very low (almost zero) repairs if we do nine or more oil changes but about $680 if we do no oil changes.

Next, let's try using both oil_changes and miles to create a new linear model. Perhaps both variables could do a better job predicting repair costs (as opposed to using only oil_changes). We can do this with the following code:

```
X = oil[['oil_changes','miles']]

y = oil['repairs']

lm = linear_model.LinearRegression()
lm.fit(X, y)
y_pred = lm.predict(X)

print('Coefficients (slope): \n', lm.coef_)
print('Intercept:', lm.intercept_)

r2 = metrics.r2_score(y, y_pred)
print('R-Squared: ', round(r2,2))
mse = metrics.mean_squared_error(y, y_pred)
print('RMSE: ', round(np.sqrt(mse),2))

Coefficients (slope):
 [-7.19859081e+01   1.50835832e-02]
Intercept:    343.2656691937482
R-Squared:    0.88
RMSE:    70.9
```

There's only on small difference between this code cell and our earlier regression calculation: In the first line of code, we include two predictor variables, namely, oil_changes and miles. The output is slightly different, too: There are two coefficients listed, and they are both expressed in scientific notation. The first coefficient says that the slope for oil_changes is negative $71.99 and the slope for miles if 0.015. We already know how to interpret the first slope. What the second slope is saying is that for every additional mile driven, the cost of repairs increases by about a penny and a half. We can also see that the new model has a slightly higher R-squared value compared to the previous model's value and the RMSE has declined by about six dollars to $70.90. This result suggests that miles driven improves the predictive capability of the model by a little bit. So, in this case, we could probably stick with our original model, where we predict repair costs as a function of the number of oil changes.

Now here's where some data science thinking comes into play. In the models we have computed so far, we've ignored one important element of the situation: We did not consider the cost of an oil change into our analysis. There's the cost of the new oil and filter, the cost of the labor to do the oil change, and the cost of disposing of the old oil in an environmentally responsible manner. Additionally, when the car is in the garage having its oil changed, no one can use the car for company business, so there is also a price tag for the lost productivity. Putting that all together, we need to model the cumulative costs of the oil changes over the service life of each vehicle. The following code cell contains an additional list of data indicating the cumulative costs of all of the oil changes accomplished on each vehicle. Using these data we can compute total costs for each car, that is, the sum of the lifetime cost of the repairs plus the cumulative cost of all the oil changes.

```
oil['oil_change_cost'] = [475,750,325,485,125,590,\
                          900,705,385,275,0,1125,925,975]

oil['total_cost'] = oil['oil_change_cost']+oil['repairs']

oil[['oil_change_cost', 'total_cost']].head(4)
```

	oil_change_cost	total_cost
0	475	775
1	750	1050
2	325	825
3	485	885

Now we can build a more realistic cost model that takes into account the fact that if you take a car off the road several times to change the oil, there are real costs associated with that preventive maintenance. The following code is similar to the previous two linear models we created:

```
X = oil['oil_changes']
X = np.array(x)
X = X.reshape(-1, 1)
y = oil['total_cost']

lm = linear_model.LinearRegression()
lm.fit(X, y)
y_pred = lm.predict(X)

print('Coefficients (slope): \n', lm.coef_)
print('Intercept: \n', lm.intercept_)
```

```
r2 = metrics.r2_score(y, y_pred)
print('R-Squared:   ', round(r2,2))
mse = metrics.mean_squared_error(y, y_pred)
print('RMSE:   ', round(np.sqrt(mse),2))

Coefficients(slope):
 [17.28932584]
Intercept:
 856.5870786516854
R-Squared:    0.11
RMSE:    133.01
```

This analysis produces a dramatically different result. In this case, each additional oil change is associated with $17.29 in additional lifetime maintenance costs (cost of oil changes plus cost of repairs). The intercept term shows that if no oil changes are accomplished, total repairs are predicted to be $856.59. Our model quality has greatly diminished with an R-squared of 0.11 and an RMSE of $133.01. This substantial decrease in quality suggests that the cumulative cost of oil changes may contain a lot of random variation perhaps because the amount of time each car spends in the shop varies.

Part of the point of this example is to sensitize you to the so-called omitted variable problem. Whenever we have data on a situation, it is virtually inevitable that we do not know the status of every possible variable. Additionally, we may neglect to include particular variables in a model or intentionally decide not to include a variable for the sake of simplicity and ease of interpretation. Before we took the materials, labor, and time costs of oil changes into account, it looked like our model was simply telling us to change the oil as much as possible. Our updated model has a more sophisticated view of the matter. Keep in mind though that even in our more sophisticated model, we have still not accounted for everything. For example, when we liquidate the fleet at the end of 3 years, the resale value of a well-maintained car will almost certainly be greater than that of a car that has never had its oil changed!

PARTITIONING INTO TRAINING AND CROSS VALIDATION DATASETS

There is one last important concept that we have not yet covered with respect to supervised learning (i.e., working on a classification or regression problem). Specifically, a predictive model, created by training an algorithm on our data, will inevitably make mistakes. For example, the model might predict rain for Thursday, but it turns out that Thursday is sunny. When predicting repair costs of $2,000, we might actually have repair costs of $2,350. Improving prediction accuracy (i.e., reducing the error rate) is the basic goal of supervised machine learning: We use our training cases to help the algorithm discover the underlying pattern. Once the algorithm has detected the pattern from the training data, it stores coefficients that describe the trained machine learning model. We can then apply the model to some new X data (i.e., new values of our predictors), as we did in the previous repair example.

As you have seen, we saw how well our model performed on new data by checking predicted values against the actual events using the original training data. That's actually a little risky: Our model optimized itself to fit the training data, so reusing the training data gave the model an unfair advantage. Through hard experience, data scientists have often found that a model rarely performs as well on new data as it does on the training data.

Here's a cool new trick, however: Rather than recycling training data for evaluating model quality, we can divide the data that we have into two chunks, a training dataset and a test dataset (which is also sometimes called a holdout sample or a cross validation sample). For example, we could randomly sample two-thirds of that data that we already have to train our model and save the rest of the data to use in *testing* our trained model. Using a weather example, if we had historical weather data from 900 different days, we might randomly sample n = 600 days for training and then save the remaining n = 300 days of data to use in *testing* our trained model.

A test dataset helps us avoid the problem of overfitting. The simple regression model we demonstrated in this chapter, by definition, finds the optimal best-fitting line for the training data. In theory, many of the more advanced machine learning algorithms, such as the techniques we will use in the next couple of chapters, are so powerful that they can memorize the input data and so can perfectly predict the outcome data in the training dataset. In this case, the overfitted model won't generalize to new data. In effect, a model may become too specialized on the training data and won't work anywhere else. In most cases, the whole point of creating a model is to use it to predict values on future data that the model has never seen. In short, we want to make sure our model generalizes from our training data to new, unseen data.

With this in mind, we generally want to have a dataset to train (i.e., build) the model and a different dataset to test (i.e., evaluate) the model. There is no universal approach on how much of the data should be set aside for training, and how much for testing, but as a rule of thumb, you can use somewhere between two-thirds to three-quarters of the dataset to train and the remainder to test. When using this partitioning approach, it is also important to randomize the selection of cases for the training and test sets to ensure that there is no systematic bias in the selection of cases.

Fortunately, the sklearn package contains the train_test_split() function, which randomly samples without replacement to create training and testing datasets. One important aspect of the train_test_split() function is the stratify parameter, which defines the attribute (variable) that should be used to ensure that the differing values of that value are equally well represented in both the training and test sets. Generally, one specifies the outcome variable as the basis of stratification, and train_test_split() makes sure that both the training and testing sets are representative of the whole dataset with respect to the distribution of the outcome variable.

USING K-FOLD CROSS VALIDATION

Partitioning your data into train and test datasets can work reasonably well if you have a lot of data—let's say more than 1,000 observations/rows. However, if you do not have a lot of data (like our car maintenance example), or if your data are complex, then partitioning your data might not be the best strategy. In addition, results might vary a lot with small datasets depending upon random factors that influenced which data ended up in the training set and which in the test set.

In 1995 a researcher named Ron Kohavi published a study that compared the partitioning strategy to a few other methods. After looking over the results, Kohavi recommended a different approach, known by the phrase "stratified k-fold cross validation." With the stratified k-fold strategy, you tell your training algorithm to follow these steps: 1) randomly divide *all* of your available data into k separate chunks (it turns out that k = 10 is often an excellent choice) where each chunk has roughly the same balance of outcomes (i.e., if it rains on average on two of every 10 days, then every chunk of training data must contain 20% rainy days); 2) have the algorithm train k different models, where each of the k chunks is set aside from the training data one time (you can use the held-back chunk to evaluate the training success); 3) combine the results of the k separate models into one overall model; 4) report the error rate and accuracy of the overall model.

The k-fold cross validation process helps us avoid overfitting, and we can feel assured that the resulting model quality will hold up in future samples. Note however, that professional data scientists often work with clients who demand a demonstration of cross validation in a holdout sample. If you have lots of data for training (e.g., thousands of observations), go along with what the client wants. There's no harm in holding out some test data if you have a lot of training data, and your results when testing predictions in the test data should come out close to what your k-fold cross validation process produced when training the model in the first place.

Once again sklearn comes to the rescue: To use k-fold cross validation, we can call the RepeatedKFold() function, stating the number of folds and the number of repeats. The repeated part of the cross validation is important in that doing k-fold cross validation typically generates different results that depend on how the data were split. So, to account for this variation, we typically repeat the whole k-fold cross validation multiple times and average all of the results together.

To summarize, in this chapter, we explored how to model relationships among variables by using a simple linear model. We looked at best-fitting lines, the meaning of R-squared, and how to interpret the slope, also known as b, value in a linear model. We also discussed how the sklearn package can partition data intro training and test sets. We also discussed how k-fold cross validation can do a better job than simple partitioning, particularly when we do not have a lot of data. To emphasize this point, we discussed the manual partitioning strategy not because it is better but because certain audiences for data science results may insist that you show them how your trained model performs on a holdout sample. We should also note that there are other training strategies besides k-fold cross validation and partitioning (e.g., bootstrapping), which may also sometimes be requested by clients. In the next chapter, we will use sklearn to explore classification machine learning algorithms.

CASE STUDY: BUILDING A LINEAR MODEL USING SURVEY DATA

```
Case Key Points:
- Build a linear model to predict Likelihood.to.recommend
- Create a multi-predictor model
- Use test_train_split() to do partitioning
- Use RepeatedKFold and compare to partitioning
```

Let's read in the survey dataset and try to build a linear model predicting Likelihood.to.recommend from Arrival.Delay.in.Minutes, Flights.Per.Year, and Age. We can use the same code as before to read in the large dataset. In addition to that, let's filter out missing data from our variables, create our matrix of X values, and instantiate an empty linear model object:

```
survey = survey.dropna(subset=['Arrival.Delay.in.Minutes'])
survey = survey.dropna(subset=['Age'])
survey = survey.dropna(subset=['Flights.Per.Year'])

X = survey[['Arrival.Delay.in.Minutes','Age',\
            'Flights.Per.Year']]

y = survey['Likelihood.to.recommend']

lm = linear_model.LinearRegression()
```

The first three lines drop missing data based on each individual predictor. We could have combined these three steps into a single line of code, but this is easier to read. The next line creates our matrix of X values. We are going to use these to create predictive models using two different data management strategies as described at the close of this chapter. First, we will partition the data into 75 percent training data and 25 percent test data. We will train the lm model on the training data and evaluate it on the test data. You may remember from previous chapters that we have more than 88,000 observations (rows) in this dataset: That's plenty for a partitioning strategy.

Then we will compare the partitioning strategy with a k-fold repeated cross validation. Even with a large dataset, k-fold cross-validation is likely to provide a more complete picture of model quality. First, here's the partitioning code:

```
from sklearn.model_selection import train_test_split

X_train, X_test, y_train, y_test = \
                train_test_split(X,y,test_size=0.25,\
                stratify=y,random_state=42)

lm.fit(X_train, y_train)
y_pred = lm.predict(X_test)

print('Coefficients (slope): \n', lm.coef_)
print('Intercept: \n', lm.intercept_)

r2 = metrics.r2_score(y_test, y_pred)
print('R-squared: ', round(r2,2))

Coefficients (slope):
 [-0.00581263 -0.02194134 -0.03086776]
Intercept:
 9.039426230711932
R-squared: 0.08
```

In the first line of code, we import the train_test_split() function from sklearn's model_selection module. Next, we run the function, supplying the X and y variables as well as the proportion of the data, 25 percent, that we want to dedicate to the test set. The final two arguments to train_test_split() specify the y variable so that we get an even distribution of different values of Y across the two data subsets and provide a random number seed to make the results reproducible. Note that train_test_split() returns a so-called tuple consisting of four separate objects, an X and Y value for the training, and an X and Y value for the test.

After that the code is as we saw earlier in the chapter, with the only difference being that when we call lm.predict(), we supply X_test as the inputs, and when we call metrics.r2_score(), we supply y_test as the dataset for making comparisons to the predicted values of Y. The net effect of this code cell is that we train the model on the training data and we evaluate the model on the test data. The output shows the three slope values, which are all negative! The longer the flight delay, the lower the LTR. The higher the age, the worse the LTR. And the more flights the passenger has taken in the past year, the lower the LTR. In a nutshell, older, more experienced flyers who experience a flight delay are relatively unhappy. The R-squared value of 0.08 is relatively small, but don't forget that when we are dealing with the unpredictability of human behavior (including survey-taking behavior), it is not unusual to have small R-squared values.

Keep those results in mind as we proceed to use repeated k-fold cross-validation. As a reminder, the K-fold process runs the analysis several times, and each time it holds back a different subset of the data to use as a validation sample. Here's the code:

```
from sklearn.model_selection import RepeatedKFold

cv_rep_KFold = RepeatedKFold(n_splits=10,
                    n_repeats=4, random_state=42)

type(cv_rep_KFold)

sklearn.model_selection._split.RepeatedKFold
```

In this code cell, we import the RepeatedKFold() function from sklearn's model_selection module. We then create an instance of the RepeatedKFold class, specifying the splitting and repeating we want the procedure to do. In this case, we use the n_splits=10 argument to conduct k = 10-fold cross validation. That signifies that we will hold back about one-tenth of the data (roughly n = 8,800 rows) to serve as the test (validation) set during each fold. We will do that process 10 times, each time holding back a different subset of the data. That's not the only repetitive thing we will be doing, however. We are also going to repeat that entire k = 10-fold process four times, which we have specified with the n_repeats=4 argument. In total, then, we will be running the linear model analysis 40 times. Each time will be producing a slightly different result because it will be using slightly different chunks of data for training and test.

The final line of code in the previous cell simply inquires with Python about the type of the KFold object we produced. Python reports that it is a custom class of the type RepeatedKFold. The other three parts of the reported name provide information about the "ancestors" of the class: Programmers sometime need to know this information for debugging. The only important point we need to remember is that we now have an object called "cv_rep_Kfold" that contains all of the specifications needed to perform our 40 repeats. So let's get to it:

```
from sklearn.model_selection import cross_val_score

scores = cross_val_score(lm, X, y, cv=cv_rep_KFold)

print('R-squared: Mean %.3f (SD %.3f)' % \
      (np.mean(scores), np.std(scores)))

R-squared: Mean 0.090 (SD 0.007)
```

Note that the first line of code in the cell imports cross_val_score(), which is the function we need to actually run the analysis. We supply the lm model that we want it to use, as well as all of the data, plus the cv_rep_Kfold from the previous code cell. What cross_val_score() will now do is use the information in cv_rep_Kfold to configure 40 different training samples, each with a custom holdout sample for testing. The function will run each of those 40 analyses using the linear modeling method stored in lm. It will record the results of each of those runs, specifically the R-squared value from each run, in a list. The output following the last line of code shows the mean value among the 40 R-squared values as well as the standard deviation.

Note that the mean R-squared is 0.09—slightly higher than what we got from the partitioning strategy. This shows the power of k-fold cross-validation. We have avoided the trap of being stuck with the results from only one training sample and one test sample and have thereby arrived at a more accurate understanding of the quality of our linear model. The standard deviation of 0.007 shows that there is a little bit of variation among the 40 different models. Some of them have R-squared values slightly larger than 0.09, and some have R-squared values that are slightly lower. None of those minor variations are likely to have any meaningful impact on our main conclusions about how age, number of flights taken per year, and arrival delays adversely affect LTR. And by using some additional functions from sklearn, we could also obtain "consensus" (mean) values for the slopes and intercepts across our 40 models if it were important to report those.

CHAPTER CHALLENGES

1. Create a bivariate scatterplot of the Departure.Delay.in.Minutes and Likelihood. to.recommend variables from the large survey dataset. Report whether the relationship between the two variables looks linear or something else.

2. Create a linear model that uses Departure.Delay.in.Minutes to predict Likelihood. to.recommend. Report the slope, intercept, R-squared value, and root mean squared error.

3. Create a bivariate scatterplot of the Departure.Delay.in.Minutes and Likelihood. to.recommend variables as before, but this time add the best-fitting line.

4. Use the predict() method to predict the value of Likelihood.to.recommend, assuming that Departure.Delay.in.Minutes is equal to 10.

5. Add the variable Arrival.Delay.in.Minutes to your prediction model. Make sure to report if the R-squared value improved when compared to the model that you created for exercise 2.

6. Continuing to use the air travel case study data, use Departure.Delay.in.Minutes to predict Arrival.Delay.in.Minutes. Explain why the results do or do not make sense.

7. You might find it interesting to explore a write-up of some ideas about predictive car maintenance. The following blog post includes many references to modeling and also demonstrates the challenge of collecting useful and actionable data: https://www.mckinsey.com/industries/automotive-and-assembly/our-insights/unlocking-the-full-life-cycle-value-from-connected-car-data.

11 CLASSIC CLASSIFIERS

LEARNING OBJECTIVES

Explain the two major types of supervised learning: regression and classification.

Apply two different classifiers to data with a binary outcome variable.

Interpret accuracy measures for classifiers.

MORE SUPERVISED LEARNING

In the previous chapter, we introduced the idea of machine learning as an algorithmic method of processing data. We differentiated between supervised learning—which emphasizes making predictions—and unsupervised learning, a set of techniques for finding patterns in unlabeled data. We also introduced linear regression, a simple prediction technique that has been around for a long time. Linear regression works with ordered numeric outcome variables—outcomes that can be measured on any kind of a scale (e.g., temperature, weight, miles per gallon). In this chapter, we introduce the other major type of supervised learning algorithm: classifiers. Classifiers use input data to make predictions about the membership of a case with respect to a set of discrete outcomes: Yes/No, True/False, Sunny/Cloudy/Rainy, Dog/Cat/Goat/Horse, and so on. Similar to linear regression, our goal is to develop a model using some training data and then evaluate and use that model to make predictions with some novel input data.

A CLASSIFICATION EXAMPLE

Let's create a supervised learning model that uses some demographic data to classify whether a person's income exceeds $50,000 a year. Insurance, credit card, and mortgage companies often use classifiers to process demographic data and make decisions. Predicting income category would be considered a classification problem because we are restricting the outcome to two discrete choices (<=$50,000 or >$50,000). You might be thinking, why don't we predict the actual value for income? First, that would be a regression problem (which we discussed last chapter), rather than a classification problem, but more importantly, the dataset we have available only has a binary variable for income: an outcome variable indicating that a person's income is either <=$50,000 or >$50,000. We extracted this dataset from the University of California, Irvine (UCI) Machine Learning repository. This repository is a collection of almost 500 datasets that is freely available as a service to the machine learning community. This specific dataset was donated by Barry Becker from an analysis of the 1994 census, and it typically is called the "Adult" dataset. You can explore the information about the dataset for yourself from the UCI web site: http://archive.ics.uci.edu/ml/datasets/Adult. As shown in Table 11.1, there are 14 predictors plus the target variable "income."

Although many of these variables are self-explanatory, some are trickier to understand. The name "fnlwgt" stands for "final weight," which is the number of units in the target population that the responding row represents. These weights are important for obtaining accurate estimates of population values, but we will ignore them for this analysis to keep things simple. The attribute education_num stands for the number of years of education in total, which is a continuous representation of the education categorical variable. The attribute relationship represents the person's role in the family. Finally, the capital_gain and capital_loss attributes are income from investment sources other than wages or salary.

TABLE 11.1 ■ List of Variables in the Adult Dataset With Category Options

Column Name	Possible Values
income	<=50K, >50K
age	Continuous
workclass	Private, Self-emp-not-inc, Self-emp-inc, Federal-gov, Local-gov, State-gov, Without-pay, Never-worked
fnlwgt	Continuous
education	Bachelors, Some-college, 11th, HS-grad, Prof-school, Assoc-acdm, Assoc-voc, 9th, 7th-8th, 12th, Masters, 1st-4th, 10th, Doctorate, 5th-6th, Preschool
education-num	Continuous
marital-status	Married-civ-spouse, Divorced, Never-married, Separated, Widowed, Married-spouse-absent, Married-AF-spouse
occupation	Tech-support, Craft-repair, Other-service, Sales, Exec-managerial, Prof-specialty, Handlers-cleaners, Machine-op-inspct, Adm-clerical, Farming-fishing, Transport-moving, Priv-house-serv, Protective-serv, Armed-Forces.
relationship	Wife, Own-child, Husband, Not-in-family, Other-relative, Unmarried.
race	White, Asian-Pac-Islander, Amer-Indian-Eskimo, Other, Black.
sex (gender)	Female, Male
capital-gain	Continuous
capital-loss	Continuous
hours-per-week	Continuous
native-country	United-States, Cambodia, England, Puerto-Rico, Canada, Germany, Outlying-US(Guam-USVI-etc), India, Japan, Greece, South, China, Cuba, Iran, Honduras, Philippines, Italy, Poland, Jamaica, Vietnam, Mexico, Portugal, Ireland, France, Dominican-Republic, Laos, Ecuador, Taiwan, Haiti, Columbia, Hungary, Guatemala, Nicaragua, Scotland, Thailand, Yugoslavia, El-Salvador, Trinadad & Tobago, Peru, Hong, Holland-Netherlands

Keeping all that in mind, we can now load the data from the UCI web site directly into a pandas DataFrame. Specifically, the dataset is located at:

http://archive.ics.uci.edu/ml/machine-learning-databases/adult/adult.data.

Because the UCI dataset does not have metadata, we will assign column names to the DataFrame:

```
url = 'https://archive.ics.uci.edu/ml/machine-learning-\
databases/adult/adult.data'

col_names = ["age", "workclass", "fnlwgt", "education",
"education_num", "marital_status", "occupation",
"relationship", "race", "sex", "capital_gain",
"capital_loss", "hours_per_week", "native_country",
"income"]

incomeDF = pd.read_csv(url, header=None, names=col_names)

incomeDF.info()
<class 'pandas.core.frame.DataFrame'>
RangeIndex: 32561 entries, 0 to 32560
Data columns (total 15 columns):
 #   Column          Non-Null Count  Dtype
---  ------          --------------  -----
 0   age             32561 non-null  int64
 1   workclass       32561 non-null  object
 2   fnlwgt          32561 non-null  int64
 3   education       32561 non-null  object
 4   education_num   32561 non-null  int64
 5   marital_status  32561 non-null  object
 6   occupation      32561 non-null  object
 7   relationship    32561 non-null  object
 8   race            32561 non-null  object
 9   sex             32561 non-null  object
 10  capital_gain    32561 non-null  int64
 11  capital_loss    32561 non-null  int64
 12  hours_per_week  32561 non-null  int64
 13  native_country  32561 non-null  object
 14  income          32561 non-null  object
dtypes: int64(6), object(9)
memory usage: 3.7+ MB
```

We have used a handy pandas method called info() to produce a list of columns and other information about the DataFrame. The output shows that the data structure has 32,561 rows and 15 columns. Let's explore the income variable more:

```
incomeDF.income.value_counts()

 <=50K   24720
 >50K     7841
Name: income, dtype: int64
```

We used the value_counts() function to summarize the income variable. Remember that we refer to a variable like this as a "binary or dichotomous variable." The output shows us that there are 24,720 people who were classified as having an income of less than 50K and 7,841 people who had an income of more than 50K. When there are substantially different amounts of cases for the different categories, that situation is known as an unbalanced dataset. Unbalanced outcome data has the potential to cause problems with the analysis, but we will ignore that problem for now.

We need to clean the data to get it ready for analysis. First, as previously noted, we will ignore the fnlwgt attribute, which we can do by removing it from the DataFrame. We also remove several other attributes that we will not use in our analysis.

```
incomeDF = incomeDF.drop(columns={'fnlwgt',
'native_country', 'relationship', 'marital_status'})

incomeDF.shape

(32561, 11)
```

Next, we need to improve our predictor variables by creating higher-level groupings for categorical variables such as education. This is known as "feature engineering," the process of applying human knowledge to the data to create better predictor variables. Feature engineering can help the machine learning algorithms create more accurate results and/or work more quickly. There are no generic rules to guide the process of feature engineering—rather the choices involved are a matter of experience and experimentation. Our overall goal is to create new categories from old ones, hopefully simplifying the situation along the way, while also maintaining or enhancing the predictive power of the resulting model.

We can also use feature engineering to address missing data: Sometimes simply coding a new attribute to indicate when something is missing can improve model training. In this dataset, missing elements are noted with a "?". We could try to replace those unknown values via a process of imputing those missing values. "Imputing" is a fancy word for using statistics to make an educated guess. For example, a simple approach could be to calculate the mean of the known values or the most commonly occurring value and use that value for all instances that are missing. We could also use more advanced techniques to impute missing values, but for the sake of simplicity, we will define a unique category for the missing data. In this way, we help ensure that we are not introducing bias into our model. If we removed rows that contained the missing data, we might be losing important information. As an example, what if the education value was missing primarily for those without a high school diploma? Leaving them out of the analysis could create bias in the resulting model by ignoring this important group.

The code that follows defines new levels for several attributes. We will make some changes to the workclass, education, marital_status, occupation, and native_country variables. In each case we will recode the original set of seven or more different options into a much smaller set of four to six options using the replace() command. We will also grab the "?" that was the indicator for missing data and recode that into a level called "unknown."

```
incomeDF = incomeDF.replace(' ', '', regex=True)

incomeDF = incomeDF.replace({'Local-gov', 'State-\
gov','Federal-gov'},'public')

incomeDF = incomeDF.replace({'Private','Self-emp-not-inc',
'Self-emp-inc'},'private')

incomeDF = incomeDF.replace({'Never-worked','Without-pay', '\?'},'other')

incomeDF = incomeDF.replace({"Preschool", "1st-4th",
"5th-6th","7th-8th", "9th","Assoc-acdm","Assoc-voc","10th",
"11th","12th", "HS-grad"},'noCollege')

incomeDF = incomeDF.replace({"Bachelors","Some-college"},
'graduate')
incomeDF = incomeDF.replace({"Masters", "Prof-school"},
'master')

incomeDF = incomeDF.replace({"Doctorate"},'phd')

incomeDF = incomeDF.replace({"Adm-clerical"}, 'clerical')

incomeDF = incomeDF.replace({"Craft-repair", "Machine-op-\
inspct","Other-service"},'tradeskill')
```

```
incomeDF = incomeDF.replace({"Sales","Tech-support","Armed-\
Forces", "Prof-specialty","Exec-managerial"},'highskill')


incomeDF = incomeDF.replace({"Farming-fishing"},'agricultr')


incomeDF = incomeDF.replace({"\?"},'unknown')

print(incomeDF.shape)
print(incomeDF.head(1))

(32561, 11)
age  workclass  education  education_num  occupation  race   sex
39    public    graduate        13          clerical  White  Male


capital_gain  capital_loss  hours_per_week  income
    2174           0             40         <=50K
```

There's a lot of punctuation in each of those calls to the replace() method. The effect is quite simple, however: The options between the curly braces are replaced by the string value that appears after those options. For example, in the second call to replace() in this code block, we ask replace() to search the whole DataFrame for 'Local-gov', 'State-gov', or 'Federal-gov' and, whenever any one of these is found, to replace it with the string 'public'. When a call to replace() has more than one option to search for, that signifies we are compressing two or more options into a single new category.

You might be wondering how we decided on the new set of options for each variable. For example, we recoded the 10 original options for workclass into public, private, and other. It is possible to use statistical analyses as a guide to useful choices. For instance, we could have compared instances where workclass was coded as Federal-gov to see if they were similar to those coded as State-gov. In this case, we used knowledge from previous projects suggesting that public sector workers share many similarities but are meaningfully different from private sector workers in several respects. There is no one correct way to do this kind of recoding— you use your knowledge of the problem domain as a starting point and also be ready to revise your choices if analytic results suggest that you should.

SUPERVISED LEARNING WITH NAÏVE BAYES

Using this dataset, let's now create a model to classify people into low or high income. The algorithm needs to use the predictors (i.e., the attributes we cleaned and engineered) to divide the data into two classes (i.e., <=50K and >50K). We will first use a technique called "Naïve Bayes analysis" to perform the classification. This technique is a common machine learning approach because of its simplicity and flexibility. The technique is called "naïve" because it makes an assumption that each predictor variable works independently of all the others—an assumption that is not always true but nonetheless usually works pretty well. Bayes refers to a 19th-century statistician named Thomas Bayes who figured out a basic form of statistical logic that is at the heart of this analysis technique.

Let's pretend for a moment that we are using one numeric predictor called "x," to divide data into two classes: income <=50K and income >50K. Let's also assume that low values of x tend to predict income <=50K and high values of x tend to predict income >50K. Put in the simplest terms, the Naïve Bayes algorithm finds the spot on the number line of x that best differentiates the <=50K and >50K cases: Values of x lower than that spot will predict income <=50K, and values of x higher than that spot will predict income >50K. Of course, we typically have more than one predictor, so what Naïve Bayes does is to combine each predictor's guess about the output class into an overall probability value. If most of the predictors are hinting at income <=50K, then that's the outcome that will have the highest probability, and that's the outcome that will be predicted (and vice versa).

NAÏVE BAYES IN PYTHON

Before we can start to build a Naïve Bayes model in Python, we need to review ways of encoding categorical variables as numbers. The Naïve Bayes technique (and most other machine learning techniques) expect their inputs to be in the form of numbers. There are several methods of encoding including "dummy coding" and "one-hot encoding." Dummy coding is preferred by statisticians because of its beneficial statistical properties. One-hot encoding is often used by computer scientists, particularly when training machine learning models, because it simplifies some programming tasks.

In dummy coding, we represent k categories as k-1 binary variables. For example, if we had three colors that we needed to dummy code—red, green, and blue—we would represent those as two binary variables, which we could call color_1 and color_2. The color_1 variable would be equal to one if the color were red and zero if it were blue or green. Then, color_2 would be equal to one if the color were green and zero otherwise. Blue would therefore be represented as a zero for color_1 and a zero for color_2.

One-hot encoding represents k categories as k binary variables. Each of the binary variables has all 0s *except* for the rows where that attribute is present, in which case the variable has the value of 1. The code examples that follow demonstrate how one categorical variable with three possible values (["Blue ", "Green ", "Red "]) gets mapped to dummy variables and one-hot encoded variables.

```
colors = pd.Series(["Blue", "Green", "Red"])

pd.get_dummies(colors, drop_first=True)
```

	Green	Red
0	0	0
1	1	0
2	0	1

Note that this dummy coding is accomplished by pandas, which conveniently, can recode an entire DataFrame in one step. The drop_first=True argument provides the classic dummy code configuration, which has one fewer variables than the number of categories. The topmost row of the output (with the 0 label) shows how the original value of "Blue" is encoded as 0 and 0. Here's the one-hot encoding of the same data:

```
colors = np.array(["Blue", "Green", "Red"])

ohe = OneHotEncoder(sparse_output=False)
ohe_fitted = ohe.fit_transform(colors.reshape(-1,1))

print(ohe_fitted)
print(ohe.categories_)

[[1.0.0.]
 [0.1.0.]
 [0.0.1.]]
[array(['Blue', 'Green', 'Red'], dtype='<U5')]
```

In this case the output shows three columns of data. The topmost row here is encoded as 1, 0, 0 as an indication of "Blue." The encoding occurred in two steps. First we instantiated OneHotEncoder, telling it that we did not want a sparse representation of the results (if set to True, this creates a compressed version of the output data that cannot be used for analysis). In the second step, we fit our tiny dataset to create the new encoding variables. Note that the OneHotEncoder is a specialized class offered by sklearn's preprocessing module. At the beginning of the notebook for this chapter, you can see a code cell where we imported that class.

Now, finally, we can create a model using Naïve. First, we will build the model using the simple partitioning approach with separate datasets for training and testing. So, as we did in the previous chapter for linear models, we will create a test and train dataset. Review the previous chapter if you need additional hints understanding this code.

```
simple_edu_df = pd.get_dummies(incomeDF, \
                               sparse=False, drop_first=True)

y = simple_edu_df['income_>50K']
```

```
X = simple_edu_df.drop(columns=['income_>50K'])

X_train, X_test, y_train, y_test = \
    train_test_split(X, y, test_size=0.3, random_state=10)

print(X_train.shape,X_test.shape,\
    y_train.shape,y_test.shape)

(22792, 24) (9769, 24) (22792,) (9769,)
```

The first line of code in this cell uses the pandas get_dummies() utility to create dummy codes for each of the categorical variables in the original dataset. Then we extract the binary y variable from the dataset (income > 50K) and the matrix of X variables, comprising all of the other columns. We use train_test_split() as we did in the previous chapter, and the results show 24 predictor variables, n=22792 cases in the training data, and n=9769 cases in the test data.

We are now ready to build our Naïve Bayes model. The following command generates a model based on the training dataset.

```
nbModel = GaussianNB() # Naive Bayes model
nbModel.fit(X_train, y_train)

GaussianNB
GaussianNB()
```

The output simply confirms that we have created a trained Naïve Bayes model. The term "Gaussian" refers to the distribution that the algorithm uses to construct the probabilities for each predictor variable. The following lines of code evaluate our model:

```
y_pred = nbModel.predict(X_test)

from sklearn.metrics import classification_report

print(classification_report(y_test, y_pred))

              precision    recall  f1-score   support

           0       0.84      0.92      0.88      7423
           1       0.63      0.43      0.51      2346
```

Accuracy			0.80	9769
macro avg	0.73	0.67	0.69	9769
weighted avg	0.79	0.80	0.79	9769

In the first line of code, we create a set of predicted y values from the test data. Then we import the classification_report() function from sklearn. This function provides a lot of useful output, but it is easiest to start in the middle with the line that begins with "accuracy." This is the overall accuracy of the model, that is, the total number of correct classifications divided by the total number of cases in the test set. The particular value we found, 0.80, must be interpreted in terms of the problem we are trying to address. As a first effort to predict income > 50K, this is a reasonably good value. Obviously, if we were making predictions about a medical condition or some other critical outcome, then 0.80 would not be considered particularly good.

There are three other qualities reported in this output: precision, recall, and F1. Like accuracy, precision and recall are also ratios, so they range from 0 to 1 and values closer to 1 are preferred. Precision is the ratio of true positive predictions (in our case, correct predictions of income > 50K) divided by the sum of the true positives and false positives. Precision is therefore a measure of accuracy that is primarily focused on our success at predicting whichever of our categories is encoded as a 1. Recall is the ratio of true positives to the sum of true positives and false negatives. This is the same measure that some scientists call "sensitivity," and it also focuses primarily on our success at predicting whichever of our categories is encoded as a 1. Finally, the F1 score (often called F) is the harmonic mean of the precision and recall values. Researchers like the F1 score because it summarizes model quality in a balanced way. When making a report of the performance of a classifier, it is advisable to report the basic accuracy measure as well as precision, recall, and F1 scores.

Note that there are several other measures of classifier quality not represented in the output of classification_report(). These include specificity, which is the true negative rate; Type I and Type II error, which are favorites of statisticians and are closely related to specificity and sensitivity; Cohen's Kappa accuracy, which accounts for unbalanced categories; and area under the receiver-operating characteristic curve, which is frequently used in signal processing applications. Some data science clients may request these or other measures, and formulas for computing them are readily available from Wikipedia and other sources.

There's one more diagnostic that is used quite frequently by data scientists to get the big picture of how well a classifier is working. This is known as the "confusion matrix," and it is a table that shows the counts of all of the possible correct and incorrect classifications. Here's the confusion matrix produced by our Naïve Bayes model using our test data:

```
print(pd.crosstab(y_test,    y_pred,   rownames=['Actual'],
                  colnames=['Predicted'],   margins=False))

Predicted      0      1
Actual
0           6826    597
1           1341   1005
```

To interpret this confusion matrix, first look at the main diagonal, which comprises all of the correct predictions. The 6,826 is the count of correct predictions of 0 (i.e., income <= 50K) and the 1,005 is the count of the correct predictions of 1 (i.e., income > 50K). We want as many of the cases from the test data to fall into one of those two categories. The off-diagonal contains the erroneous predictions: There are 1,341 false negatives (predicted as 0 but actually 1) and 597 false positives (predicted as 1 but actually 0). We want as few of the cases from the test data to fall into these two categories. This confusion matrix gives a quick overview of the classifier's performance: For example, we can see that there are more than twice as many false negatives as there are false positives. This is quite a common occurrence when using a dataset that is unbalanced in favor of the 0 category.

Keep in mind that we are sometimes more interested in one kind of error rather than the other kind. Consider which is worse: a person who gets mistakenly classified as having income >50K or a person that gets mistakenly classified as having income <=50K. Perhaps these two types of errors are equivalent, but in some situations, one might be worse than the other. For example, if we are trying to identify potential students for need-based financial aid so that we could send those students information on financial aid, we might be more concerned about being mistakenly classifying someone as having more income than they actually have (a false positive in terms of the model we created) because, in this situation, someone who needs information on how to apply for needs-based financial aid would not get the information. The other kind of error, where someone who does not qualify for aid nonetheless gets an information packet might not be a big deal because they could ignore it.

So now let's try k-fold cross validation.

```
from sklearn.metrics import classification_report,\
                            accuracy_score, make_scorer
from sklearn.model_selection import RepeatedKFold,\
                            cross_val_score

originalclass    =    []
predictedclass   =    []

def my_classification_report(y_true, y_pred):
    originalclass.extend (y_true)
    predictedclass.extend (y_pred)
```

```
    return    1

cv = RepeatedKFold(n_splits=10, n_repeats=4,\
                   random_state=100)

cross_val_score(nbModel, X=X, y=y, cv=cv,\
            scoring=make_scorer(my_classification_report))

print(classification_report(originalclass, predictedclass))
```

	Precision	recall	f1-score	support
0	0.84	0.92	0.88	98880
1	0.65	0.44	0.52	31364
accuracy			0.81	130244
macro avg	0.74	0.68	0.70	130244
weighted avg	0.79	0.81	0.79	130244

This code is similar to what we used in the previous chapter for k-fold repeated cross validation, but we have also added a custom function called "my_classification_report()." This function creates two arrays (one array for the original (actual) values and the other array for the predicted values across all 40 of our cross validation runs. Our new function is used in the cross_val_score() function. In the object that we use to control the training process (cv), we defined the number of folds to be 10 with four repetitions for a total of 40 distinct model results. The output now shows an overall accuracy of 0.81—slightly better than the results we obtained from partitioning. As before, our lowest model quality scores—precision, recall, and F1—are obtained when we are trying to predict the income>50K category.

SUPERVISED LEARNING USING CLASSIFICATION AND REGRESSION TREES

Naïve Bayes analysis has the advantage of simplicity: Each predictor variable (attribute) is treated independently. Although it is possible to add more diagnostics to the output, we don't have any sense of how the predictor variables work together or which variables contribute the most influence to making successful classifications. In the examples that follow, we will explore a different machine learning approach that can provide a more intuitive explanation of how the model works. When building machine learning models, you need to think about whether model transparency is important for your project. For some data science clients, the most important aspect of the analysis is being able to explain it to an intelligent person who does not have any technical know-how.

Our new approach, Classification and Regression Trees, can be imagined as a procedure similar to the game 20 Questions. We ask a series of questions, and the answer to each subsequent question helps further reduce the uncertainty about which classification is most likely correct. In this example, we will use sklearn's DecisionTreeClassifier class, which is one of several available classification tree methods offered by sklearn. In our example, we will define a decision tree of rules that defines whether we predict that a person has income <=50K or >50K.

In the code cell that follows, we define a new function "create_one_hot" to do the one-hot encoding for the categorical features. This is an alternative approach to the pandas dummy coder that we used in the previous example. After creating the one-hot encoding, we combine the different DataFrames via the concat command.

```
y = incomeDF.income

features = ['age', 'education', 'workclass', 'race', 'sex']
X = incomeDF[features]
X.reset_index(drop=True, inplace=True)

def create_one_hot(col_name):
    ohe = OneHotEncoder(handle_unknown='ignore')
    encoded_info = ohe.fit_transform(X[[col_name]]).toarray()
    enc_df = pd.DataFrame(encoded_info)
    enc_df.columns =ohe.get_feature_names_out()
    del X[col_name]
    return enc_df

enc_workclass_df = create_one_hot('workclass')
enc_edu_df = create_one_hot('education')
enc_race_df = create_one_hot('race')
enc_sex_df = create_one_hot('sex')

X = pd.concat([X, enc_workclass_df, enc_race_df,\
               enc_sex_df, enc_edu_df], axis=1)
print(X.shape)
X.columns

(32561, 16)
Index(['age', 'workclass_?', 'workclass_other',
'workclass_private', 'workclass_public', 'race_Amer-Indian-
Eskimo', 'race_Asian-Pac-Islander', 'race_Black',
'race_Other', 'race_White', 'sex_Female', 'sex_Male',
'education_graduate', 'education_master',
'education_noCollege', 'education_phd'], dtype='object')
```

This output confirms that we have 32,561 rows and 16 columns in our predictor matrix. The 16 columns represent one column for age, which is numeric, plus the one-hot encoding variables for the five original variables we selected—age, education, workclass, race, and sex. Remember that one-hot encoding creates one new column for each option in the original variable.

The code for using DecisionTreeClassifier is similar to the code we used to create the Naïve Bayes model. Notice that in this example, we use the partitioning approach to create one training dataset and one testing dataset.

```
X_train, X_test, y_train, y_test = \
    train_test_split(X, y, test_size=0.3, random_state=10)

dtree = DecisionTreeClassifier(random_state=10)

dtree.fit(X_train, y_train)
y_pred = dtree.predict(X_test)

accuracy = metrics.accuracy_score(y_test,y_pred)*100
print(accuracy)
print(pd.crosstab(y_test, y_pred, rownames=['Actual'],
           colnames=['Predicted'], margins=False))

77.8482956290306
Predicted    <=50K    >50K
Actual
<=50K         6720     703
>50K          1461     885
```

We can see in the first line of output that the accuracy we obtained (~78%) was a bit less than our model built using Naïve Bayes mainly because we used fewer predictors. However, because we are creating decision trees, we can visualize the trees directly using the following code:

```
class_names=['<=50K', '>50K']
fig, ax = plt.subplots(figsize=(16, 10))
tree.plot_tree(dtree,
            feature_names = list(X_train.columns),
            class_names=class_names,
            filled = True,
            fontsize=12,
            max_depth=2,
            ax=ax)
plt.savefig('11-dt_1.pdf')
```

FIGURE 11.1 ■ **Decision Tree for Classifying Income<=50K or Income>50K**

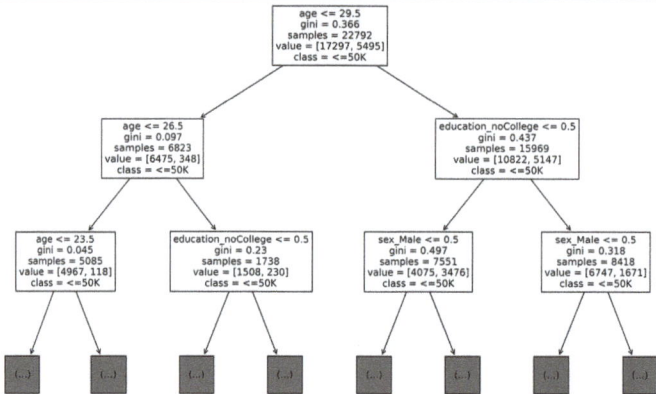

This code produces the tree plot shown in Figure 11.1. Note that the overall shape of the diagram provides a flavor of how the model works. The top-level split is age, where people younger than or equal to than 29.5 go into the left subtree, and people older than 29.5 go into the right subtree.

To understand this graphic, we will explore the first node in the right subtree, which starts with education_noCollege<=0.5:

- The first line is the attribute, in this case the encoding of whether the person attended college. Note that in the dataset, this attribute has the value of 0 or 1, so the conditional of <= 0.5 splits the no college to the left and yes college to the right.

- Gini is a metric, also known as "impurity," which ranges from 0 to 0.5. Values closer to 0 indicate that a node strongly favors one or the other class. Technically, Gini represents the probability of a future random instance being misclassified at this node of the decision tree.

- Samples notes the total number of instances (rows) being considered at this part of the tree.

- Value shows the number of cases being assigned by this node to the left or right subtree below this node. These two numbers always sum to the number of samples, as noted in the previous item.

- Class shows which attribute is represented by the left subtree underneath this node.

Putting all of this information together, this particular node places n=10,822 cases in the income<=50K category and n=5,147 cases in the income>50K category based on whether the person in question attended college. Of course, you can't read this node in isolation because the node above it (the root node) already did a split based on age, and the two nodes below it both did a split based on sex_Male. Nonetheless, we already have a sense of which variables matter in making a classification.

We can get an even clearer picture of which variables matter by exploring the importance scores assigned to each attribute:

```
df = pd.DataFrame({'feature':X.columns,\
                   'importance':dtree.feature_importances_})
df.sort_values(by=['importance'], ascending=False,\
               inplace=True)
print(df.head())

        feature              importance
0       age                  0.461569
14      education_noCollege  0.209374
11      sex_Male             0.150423
12      education_graduate   0.063955
9       race_White           0.020016
```

The first line of code in this above extracts some information from the feature_importances_ attribute that is stored on the fitted model. These values represent the accumulated impurity associated with each attribute. A higher number represents an attribute that is more powerful at making splits. In the output shown, age is the most important splitter, followed by education_noCollege, which we have discussed in some detail. Remember that there are 16 attributes in all, so by using the head() method on our small DataFrame, we have limited our view to the five most important attributes. We can also see that the attributes at the top of the tree are the most important, which makes sense because they were selected by the algorithm to be the primary drivers in splitting the tree.

We could stop here except for one important issue: Do we need all 16 of these attributes? Because a model like this could be used within the banking industry to decide which people should get more credit, it would be inappropriate (and in some places illegal) to use sex, age, and race attributes within our model. So, let's remove those attributes from our dataset and then rerun the analysis.

```
y = incomeDF.income

features = ['workclass', 'education_num', 'occupation']
X = incomeDF[features]
X.reset_index(drop=True, inplace=True)

enc_workclass_df = create_one_hot('workclass')
enc_occ_df = create_one_hot('occupation')

X = pd.concat([X, enc_workclass_df, enc_occ_df], axis=1)

X_train, X_test, y_train, y_test = \
```

```
train_test_split(X, y, test_size=0.3, random_state=10)

dtree = DecisionTreeClassifier(random_state=10)
dtree.fit(X_train, y_train)
y_pred = dtree.predict(X_test)

accuracy = metrics.accuracy_score(y_test,y_pred)*100
print(accuracy)
print(pd.crosstab(y_test, y_pred, rownames=['Actual'],\
            colnames=['Predicted']))

78.22704473334015
Predicted      <=50K      >50K
Actual
<=50K          7152       271
>50K           1856       490
```

We can see that the accuracy is slightly higher than before, suggesting that we have not lost any predictive power by removing age, sex, and race from the model and substituting workclass and occupation. The notebook for this chapter contains both a plot of the new decision tree and a list of the most important attributes: Education turns out to be the most important attribute. In short, by removing the variables that would not be ethical to use in loan or credit decisions, we have a model that is explainable and has adequate accuracy.

This completes our focus on predictive modeling. In the previous chapter we used linear regression for predictive modeling of ordered numeric variables. In this chapter, we have discussed two supervised predictive classification algorithms: Naïve Bayes and decision trees. There are many other algorithms to explore as well numerous tuning parameters and alternative training methods. The sklearn package provides access to a host of these options. Alas, the exploration of all these other possibilities is beyond our scope—but feel free to explore!

CASE STUDY: BUILDING SUPERVISED MODELS FROM THE SURVEY

Case Key Points:
- Use decision trees to predict detractors
- Explore how these results could improve likelihood to recommend

We can now generate some machine learning models that can predict whether an airline customer will be a detractor. If we could do this prediction before the flight took off, we might be able to get the airline to proactively reach out to those customers to try to improve their likelihood to recommend the airline. The notebook for this chapter contains the same code as before to read in the survey data and do basic data cleaning.

As before, we can create our train and test datasets using train_test_split(). Like our previous usage, to make sure we can reproduce our results, we used the random_state parameter to control the sequence of random numbers used to partition our data.

```
features = ['Shopped_at_airport', 'Age',
            'Airline.Status', 'Type.of.Travel']
X = survey[features]
y = survey.Detractor

X_train, X_test, y_train, y_test = \
    train_test_split(X, y, test_size=0.25, random_state=11)
X_train.shape, X_test.shape, y_train.shape, y_test.shape

((66072, 4), (22024, 4), (66072,), (22024,))
```

Next, we can build our decision tree model:

```
y = survey.Detractor

features = ['Shopped_at_airport', 'Age',
            'Airline.Status', 'Type.of.Travel']

X = survey[features]
X.reset_index(drop=True, inplace=True)

enc_shop_df = create_one_hot('Shopped_at_airport')
enc_status_df = create_one_hot('Airline.Status')
enc_type_df = create_one_hot('Type.of.Travel')

X = pd.concat([X, enc_shop_df, enc_status_df, \
               enc_type_df], axis=1)

X_train, X_test, y_train, y_test = \
    train_test_split(X, y, test_size=0.3, random_state=10)

dtree = DecisionTreeClassifier(random_state=10)
```

```
dtree.fit(X_train, y_train)
y_pred = dtree.predict(X_test)

print(classification_report(y_test, y_pred))

print(pd.crosstab(y_test, y_pred, \
            rownames=['Actual'], colnames=['Predicted']))
```

	precision	recall	f1-score	support
Detractor=False	0.85	0.89	0.87	18564
Detractor=True	0.71	0.63	0.67	7865
accuracy			0.81	26429
macro avg	0.78	0.76	0.77	26429
weighted avg	0.81	0.81	0.81	26429

Predicted	Detractor=False	Detractor=True
Actual		
Detractor=False	16589	1975
Detractor=True	2936	4929

    This code cell contains little that is new when compared with earlier examples in this chapter except for the feature names that we decided to include—Shopped_at_airport, Age, Airline.Status, and Type.of.Travel. There was no particular reason to choose these four predictors except to show off our new ability to one-hot encode the shopped, status, and type variables.

    Overall accuracy was indicated as 0.81, and the precision, recall, and F1 scores suggest that the model has an easier time predicting the Detractor=False category. That's also easy to spot from the confusion matrix that appears in the last four lines of the output: n=16589 cases were correctly classified as Detractor=False. The most common kind of error was a false negative, where the model predicted n=2,936 cases as Detractor=False when the reality was that these were Detractor=True. This highlights an issue that we ran across before: The outcome categories are imbalanced, with n=18,564 detractors but only 7,865 promoters. When categories are unbalanced like this, the model can achieve a kind of hollow success simply by overpredicting the majority category, which in this case is Detractor=False.

    Let's review variable importance to see which variables contributed the most predictive power to our decision tree classifier:

```
df = pd.DataFrame({'feature':list(X_train.columns), \
                    'importance':dtree.feature_importances_})
df.sort_values(by=['importance'], ascending=False, \
```

```
                    inplace=True)
print(df.head())

                         feature             importance
9   Type.of.Travel_Personal Travel           0.741795
6   Airline.Status_Silver                    0.213982
0   Age                                      0.024194
4   Airline.Status_Gold                      0.009428
7   Type.of.Travel_Business travel           0.003580
```

This code is the same as we used earlier in the chapter, tapping into the feature importance attribute on the fitted tree model. The most important splitter in our model is whether a passenger was doing business travel. The second most important splitter was whether the passenger had silver status on their frequent flier account. We can now use the tree visualization to understand more specifically how these splitters did their jobs. Here's the code:

```
class_names=['Detractor=False', 'Detractor=True']

fig, ax = plt.subplots(figsize=(16, 10))
tree.plot_tree(dtree,
               feature_names = list(X_train.columns),
               class_names=class_names,
               filled = False,
               fontsize=12,
               max_depth=2,
               ax=ax)
plt.savefig('11-dt_2.pdf')
```

This code produces the tree diagram shown in Figure 11.2. As we would expect from reading the feature importance output, the top of the tree shows a binary split on type of personal travel with nonpersonal travel going to the left subtree and personal travel going to the right subtree.

If the type of travel is not personal (i.e. personal travel <= 0.5), we move left, and the algorithm predicts that Detractor=Falso. This will occur in 70 percent of all surveys (n=43,522 out of the total number of samples n=61,667. In contrast, if the travel is personal, we move down to the right subtree. In either case, the next level split is on whether the traveler had silver status, but take note of how those two mid-level nodes differ: On the middle left, if a passenger does not have silver status, we predict Detractor=False and move left. On the

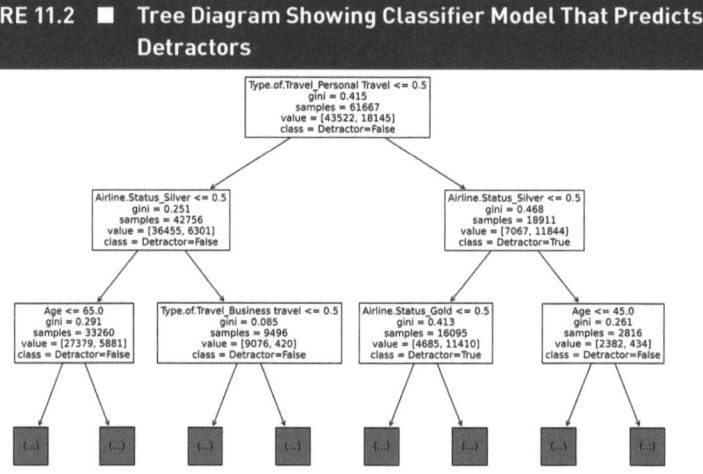

FIGURE 11.2 ■ **Tree Diagram Showing Classifier Model That Predicts Detractors**

middle right, if a passenger does not have silver status, we predict Detractor=True and move down and left to a new splitter that tests whether a passenger has gold status. We have learned something interesting and unexpected: Silver frequent flier status has a different impact depending on whether the person was traveling for personal reasons (as opposed to business or miles travel). Although it takes a bit of practice to read these trees, that is the kind of insight that data science clients may find valuable and that you will find easier to explain with the assistance of a decision tree diagram.

CHAPTER CHALLENGES

1. Load and preprocess the income dataset as shown in this chapter. Split the data into training and test sets with 50 percent for each. Run a Naïve Bayes model *without* k-fold cross validation. Compare the results to the first run in this chapter.

2. Load and preprocess the income dataset as shown in this chapter. Split the data into training and test sets with 50 percent for each. Run a decision tree model *without* k-fold cross validation. Compare the results to the first run in this chapter.

3. Extract the variables capital_gain and capital_loss from the income dataset. Run Naïve Bayes and decision tree models on these (to classify income), *without* k-fold cross validation, and compare the results to the first run in this chapter.

4. Load and preprocess the survey dataset as shown in this chapter. Create a new binary outcome variable called Promoter, like this: survey[Promoter] = survey['Likelihood. to.recommend'] >8. Then run a classification model to classify promoters, using the same four predictors as shown in this chapter, using Naïve Bayes using k-fold cross-validation with k=10 and four repeats. Compare the results to those shown in this chapter.

5. Repeat the previous exercise using decision trees instead of Naïve Bayes.

6. Using the survey dataset, add Big.Delay as a predictor to the classification model that you ran in the previous exercise.

7. Look up the term "confusion matrix" as well as Type I error, Type II error, sensitivity, and specificity. Describe how Type I error and Type II error are related to sensitivity and specificity.

12 LEFT UNSUPERVISED

LEARNING OBJECTIVES

Differentiate between unsupervised analysis and supervised analysis.

Practice a four-step analytic process: data preparation, exploratory data analysis, model development, and interpretation of results.

Use the association rules mining algorithm to uncover patterns in data.

SUPERVISED VERSUS UNSUPERVISED

In the previous two chapters, we examined predictive techniques—situations where we have an outcome variable that we want to predict. We used regression to predict a numeric variable and classification to predict a categorical variable. Data scientists call these "supervised techniques" because the outcome variable "supervises" the process of training the algorithm. In this chapter, we examine a family of analytic techniques where the algorithm has to figure out patterns in the data without having an outcome variable to guide it. We call these "unsupervised data mining" or "machine learning techniques."

One funny (but probably fictional) story about unsupervised data mining that gets mentioned quite frequently is the legendary supermarket manager who analyzed patterns of purchasing behavior and found that diapers and beer were often purchased together. The story concludes when the manager decided to put a beer display close to the diaper aisle and supposedly sold more of both products as a result. Another familiar example that online vendors use every day appears when the website says, "People who bought this book were also interested in these other books." This example is often referred to as a recommender system. Recommender systems are guided by an unsupervised data analysis technique that we will examine in this chapter. Keep in mind that machine learning is not exactly the same thing as data mining and vice versa.

DATA MINING PROCESSES

"Data mining" is a term that has been around for many years and refers to any method of discovering patterns in a body of data. Data mining typically includes four processes: (1) data preparation, (2) exploratory data analysis, (3) model development, and (4) interpretation of results. Although this sounds like a neat, linear set of steps, there is often a lot of back and forth through these processes and especially among the first three. The other point that is interesting about these four steps is that Steps 3 and 4 seem like the most fun, but Step 1 usually takes the most time. Step 1 involves making sure that the data are organized in the right way, that missing data fields are filled in, that inaccurate data are located and repaired or deleted, and that data are recoded as necessary to make them amenable to the kind of analysis we have in mind.

Step 2—exploratory data analysis—is similar to activities we have done earlier in this book: getting to know the data using histograms and other visualization tools and looking for preliminary hints that will guide our model choice. The exploration process also involves figuring out the right values for key parameters that will guide the analytic process. We will see some of that activity in this chapter.

Step 3—choosing and developing a model—is by far the most complex and interesting of the activities of a data miner. It is here where you test a selection of the most appropriate data mining techniques. Depending on the structure of a dataset, there could be dozens of options, and choosing the most promising one has as much art in it as science. We had some practice performing model development in the last two chapters, where we developed regression models and classifier models.

For the current chapter, we have decided to only explore association rules mining. So, we will not have to do Step 3 because we will not have two or more different mining techniques to compare. Remember that in the previous chapter, we did consider two different machine learning approaches to classification.

Step 4—the interpretation of results—focuses on making sense out of what the data mining algorithm has produced. This is the most important step from the perspective of the data user because this is where an actionable conclusion is formed. When we discussed the story of beer and diapers, the interpretation of the association rules guided the fictional managers choices about how to arrange product displays in the store. Translating association rules into a new configuration of store displays demonstrates the idea that useful analytic output can lead to a practical action on the part of the data user.

ASSOCIATION RULES DATA

Let's begin by talking a little about association rules. Take a look at Figure12.1 with all of the boxes and arrows:

From the arrangement of arrows in Figure 12.1, you can see that each supermarket customer has a grocery cart that contains several items from among the larger set of items that the grocery store stocks. The association rules algorithm (also sometimes called "affinity analysis" or "market basket analysis") tests out many if-then propositions, such as "If diapers are purchased, then beer is

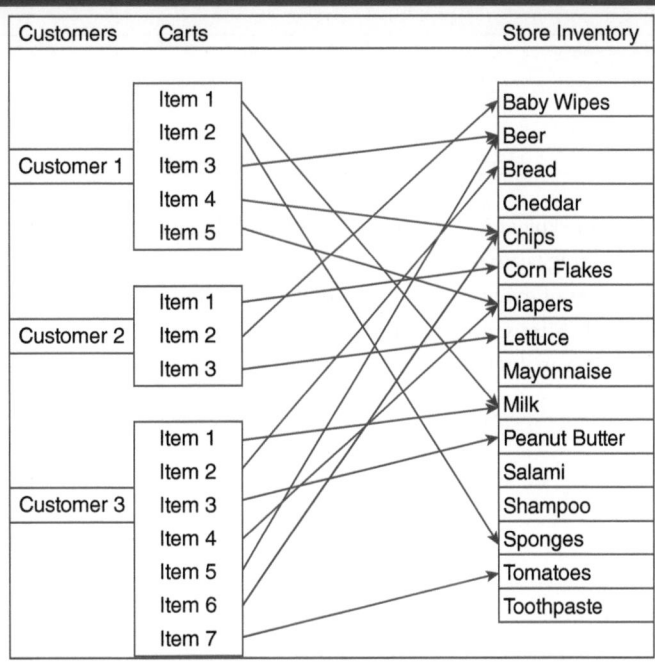

FIGURE 12.1 ■ Illustration of Association Rules With Three Shopping Carts

Note. From *An Introduction to Data Science*, by J. S. Saltz and J. M. Stanton, p. 329. Copyright 2017 by Sage Publications. Reproduced with permission.

also purchased." The algorithm uses a dataset of transactions (in the preceding example, these are the individual carts) to evaluate a long list of these rules for a value known as "support." Support is the proportion of times that a particular combination occurs across all shopping carts. The algorithm also evaluates another quantity called "confidence," which is how frequently a particular if-then rule occurs among the instances when the if item is present. In the output from an association rule algorithm, the if side of the rule is referred to as the "antecedent," and the then side of the rule is referred to as the "consequent." The consequent is generally one item, for example, beer. The antecedent, however, can have more than one item in it. For example, we might have diapers and baby food together as the antecedent with beer as the consequent.

If you look at Figure 12.1, we had support of 0.67 (the diapers–beer association occurred in two out of the three carts) and confidence of 1.0 ("beer" occurred 100% of the time with "diapers"). In practice, both support and confidence are generally much lower than in this example, but even a rule with low support and smallish confidence might reveal purchasing patterns that grocery store managers could use to guide pricing, coupon offers, shelf placement, or advertising strategies.

ASSOCIATION RULES MINING

We can get started with association rules mining easily using modules from the Python package mlxtend. In the notebook for this chapter, we retrieve the necessary packages and functions using the following commands:

```python
from mlxtend.frequent_patterns import apriori
from mlxtend.frequent_patterns import association_rules
from mlxtend.frequent_patterns import fpgrowth
from mlxtend.preprocessing import TransactionEncoder
```

We will begin our exploration of association rules mining using a dataset that is commonly available for data mining tasks: the Groceries data. Note that by using this Groceries dataset, we have relieved ourselves of much of the burden of data preparation because the dataset is nearly ready to be analyzed. After setting up the data, we move right to Step 2 in our four-step process—exploratory data analysis. You can import the Groceries dataset with this code:

```python
!git clone https://github.com/jmstanto/IDSpython.git
import os
os.chdir('IDSpython')
os.getcwd()

with open("Groceries.csv", "r") as Groceries:
    data_lines = Groceries.readlines()

import re
item_matcher = re.compile(r"\"(.+?)\"")
```

```
baskets = []

for line in data_lines:
    this_list = item_matcher.findall(line)
    baskets.append(this_list)

len(baskets)

9835
```

The first line imports a whole group of data files from a public, web-based repository. The next two lines open the Groceries.csv: This filename is unhelpful insofar as these data are not comma separated. Instead, each line has a set of grocery items in double quotes. We create a regular expression to match any quoted string and then within the for loop, we use the findall() method to produce a list of matches from each line of data in the file. Each item in the resulting list, called "baskets," is itself a list: That inner list contains a set of text strings, one to represent each grocery item on a given line of the input file. The output from this cell reports that we have n=9,835 separate baskets.

Note that the nested list we have created is a bit different than the usual rectangular dataset that we got used to earlier in the book. Let's take a look at the first three elements of the baskets list:

```
baskets[:3]

[['citrus fruit', 'semi-finished bread', 'margarine', 'ready soups'],
 ['tropical fruit', 'yogurt', 'coffee'],
 ['whole milk']]
```

The first element in baskets, shown here starting with 'citrus fruit', is a list of four grocery items, but the second element in our list has only three items and the third element has only one ('whole milk'). Happily, the folks who created the mlxtend package created a function called TransactionEncoder() to process lists of lists and make a data structure that enables the association rules analysis process.

```
te = TransactionEncoder()  # Instantiate an encoder
te_groceries = te.fit(baskets).transform(baskets)
te_groceries

array([[False, False, False, ..., False, False, False],
       [False, False, False, ..., False,  True, False],
        False, False, False, ...,  True, False, False],
       ...,
```

```
    [False, False, False, ..., False, True, False],
    [False, False, False, ..., False, False, False],
    [False, False, False, ..., False, False, False]])
```

In the first step, we instantiate a transaction encoder, and in the second step we use it to process our list of lists into a transactions object. This object is indicated as "array," which in this case is a two-dimensional structure made up entirely of True and False values. Now every possible grocery item has its own column, and the rows show the value True only in the columns where that item appears in a shopper's basket. This is not easy to see in this structure because there are no labels, so let's take a closer look.

```
te_groceries.shape # Should be 169 unique items

(9835, 169)
```

The shape method provides an indication of the extent of rows and columns: a rectangular data structure with n=9,835 rows and 169 columns, where each row is a list of items that has been added to a particular grocery cart. The columns are the individual items that could be in the cart. We can show the labels that the transaction encoder has applied to the items by using the columns_ method.

```
te.columns_[1:10] # Look at the first few items

    ['UHT-milk',
     'abrasive cleaner',
     'artif. sweetener',
     'baby cosmetics',
     'baby food',
     'bags',
     'baking powder',
     'bathroom cleaner',
     'beef']
```

We can also find out which items occur in grocery carts most frequently. If you like working with spreadsheets, you could imagine going to the bottom of the column that is marked "whole milk" and putting in a formula to sum up all of the ones in that column. You would come up with 2,513, indicating that about 26 percent of the 9,835 grocery carts contain whole milk. Remember that every row/cart that has a True in the whole milk column contains whole milk, whereas every row/cart with a False does not contain whole milk. You might wonder what the data field would look like if a grocery cart contained two gallons of whole milk. For the present data mining exercise, we ignore that problem by assuming that any non-zero number of whole milk bottles is represented by a True. Other data mining techniques could take advantage of knowing the exact amount of a product, but the association rules algorithm does not need to know that amount—only whether the product is present or absent in the cart.

Another way of inspecting our sparse matrix is by showing a graph of the frequency of occurrence of items across all carts. We can produce a bar graph to visualize this similar in approach to a histogram: It shows the frequency of occurrence of different items in the matrix. We will specify the minimum level of support needed to include an item in the plot so that we do not have to look at all 169 items at once. Remember the mention of support earlier in the chapter—in this case it simply refers to the relative frequency of occurrence of something. Let's set the support parameter to somewhere around 10–15 percent to get a manageable number of items:

```
groc_df = pd.DataFrame(te_groceries, columns=te.columns_)
freq_items = fpgrowth(groc_df, min_support=0.1,\
                use_colnames=True)
ax = freq_items.sort_values(by='support',\
        ascending=False).plot(kind='bar', x='itemsets', rot=45)
fig = ax.get_figure()
fig.savefig('Figure12_2.pdf', dpi=300, bbox_inches='tight')
```

At the start of this code cell, we dump te_groceries into a pandas DataFrame to make it easier to handle. Then we use the fpgrowth() function provided by the mlextend package to organize and filter our DataFrame. In particular, the min_support=0.1 argument ensures that we will only view items that appear in at least 10 percent of all of the baskets in the dataset. The middle line of code, starting with "ax=," strings together a couple of different commands. First, it sorts the items based on descending level of support, and then it makes a bar plot of the result. This code produces the image shown in Figure 12.2.

Figure 12.2 shows that milk is the most common item and that yogurt appears in about 14 percent of all carts. We also see a few other items such as bottled water and tropical fruit. You can modify the code to experiment with different levels of minimum support. In particular, you should test some lower values of minimum support (such as 0.05) to get a feel for how many items appear at the lower frequencies. Graphs like these help us estimate a value of support that will later give us a reasonable number of items that could potentially become part of a rule.

An item that occurs extremely rarely in the grocery baskets is unlikely to be of much use to us in terms of creating meaningful rules. Let's pretend, for example, that the item "Venezuelan Anteater Cheese" occurred only once in the whole set of 9,835 carts. Even if we did end up with a rule about this item, it would rarely apply and is therefore unlikely to be useful to store managers or anyone else. So, we want to focus our attention on items that occur

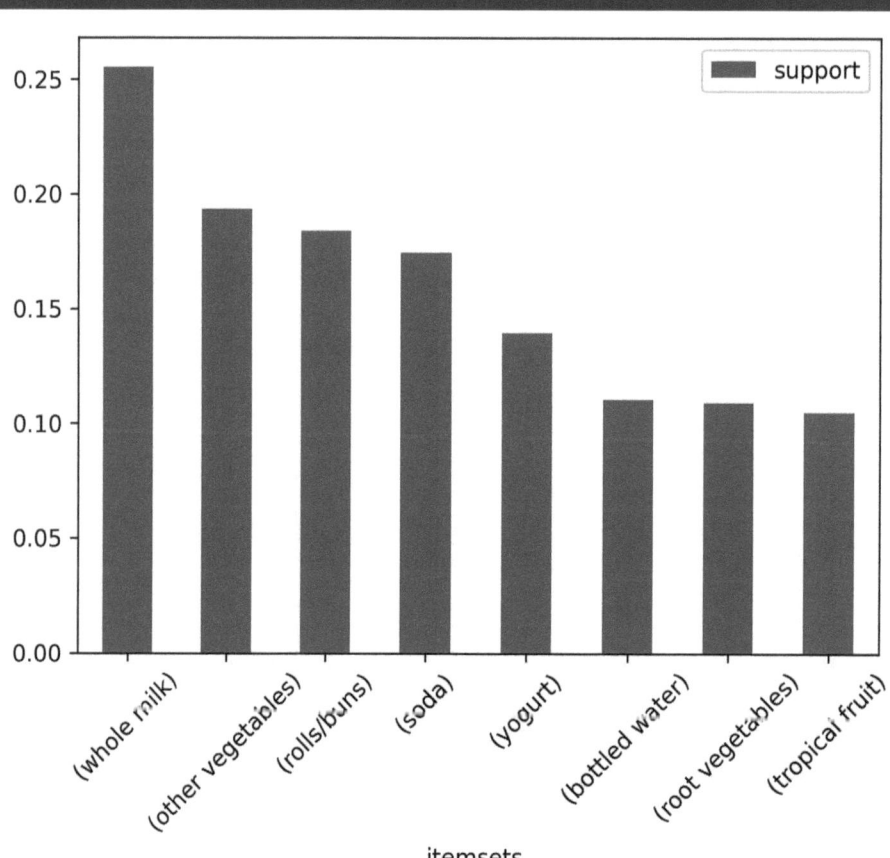

FIGURE 12.2 ■ **Support for Grocery Items With a Minimum Value of 0.10**

within a range of meaningful frequencies in the dataset. Whether this range includes 10 percent, 1 percent, 0.5 percent, 0.1 percent, or something even smaller will depend on the size of the dataset and the intended use of the rules.

Before we generate our first set of rules, let's explore one other data-related issue. In some situations, we might receive a regular rectangular dataset instead of a list of baskets, such as the grocery cart we processed. Typically, we would be interested in using association rules on such a dataset if it contained a lot of categorical items. We may be able to convert a regular rectangular dataset with categorical variables into a transactions dataset depending on the configuration of the data. Let's look at a familiar rectangular dataset and convert it to a transactions dataset. We will use the Adult dataset that we used in Chapter 11 for our discussion of classification.

Let's load up the dataset into a pandas DataFrame so that we can take a look at it again. We will need to convert each variable into separate columns representing each possible attribute—similar to one-hot encoding. After doing these transformations, we will be able to convert the pandas DataFrame into a transactions dataset, which we can then use for association rules analysis. The following code reads the data and selects seven of the columns.

```
AdultUCI = pd.read_csv('AdultUCI.csv')

AdultUCI = AdultUCI[['workclass','education',
           'marital-status','race', 'gender',
           'native-country','income']]

print(AdultUCI.columns.values)

counts = AdultUCI['education'].value_counts()
most_occuring = counts[counts>1000]
ax = most_occuring.plot(kind='bar', rot=45)
fig = ax.get_figure()
fig.savefig('Figure12_3.pdf', dpi=300, bbox_inches='tight')

['workclass' 'education' 'marital-status' 'race'
 'gender' 'native-country' 'income']
```

We eliminated columns containing widely variable numeric data, like capital-gain, because these would be tricky to recode as categorical variables. For each column that we expect to use, we should take a close look at the values or categories it contains so that we can anticipate what it will look like when converted to transactions. The last three lines of code in this cell obtain a list of counts of the values for the education column and then filters this list down to only those values that appear at least 1,000 times. The bar plot shown in Figure 12.3 reports how frequently each education category occurs.

If you image each educational category as something that a person can have "in their basket," it becomes clear how the education variable could be converted into a transactions dataset. Let's do this in two stages. First, we will go through the pandas DataFrame one time and convert the numbers to strings (as well as replacing dashes with underscores in the existing strings—this will avoid later problems with the transaction encoder). Then we will go through it a second time to make each row of data into a list of attributes that the transaction encoder can work with. Keep in mind that like all of the items in a grocery cart, each element must have a unique label. Here's the code that creates our list of attribute lists:

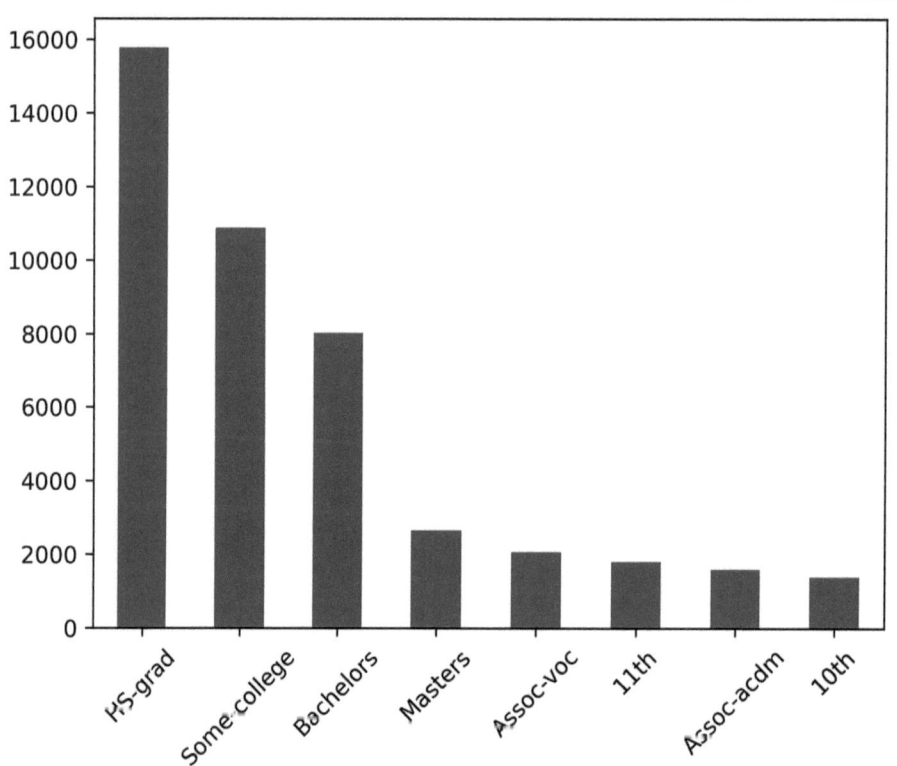

FIGURE 12.3 ■ Most Frequently Occurring Categories for the Education Variable

```
df = AdultUCI.dropna()

for col in df.columns:
    if df[col].dtype == object:
        df[col] = df[col].str.replace('-', '_')
    else:
        df[col] = df[col].astype(str)

array_values = df.values

adult_UCI_list = [list(x) for x in array_values ]
print(type(adult_UCI_list))
adult_UCI_list[0]

['Private','11th','Never_married','Black','Male','United_
States','<=50K']
```

The key to this code cell is the for loop, which steps through each of the columns in the pandas DataFrame, making one of two possible changes. If the data column is already a string (df[col].dtype == object), we simply substitute underscores in place of dashes. If the data column is numeric, we convert the numeric value into the corresponding string. The output from this code block shows the first entry in the resulting list: This is like a "basket" of attributes that the person in question possesses.

HOW THE ASSOCIATION RULES ALGORITHM WORKS

Now we are ready to convert this nested list into a transactions data set and use that to generate some rules with the apriori() and association_rules() functions. The term "apriori" refers to the specific algorithm that Python will use to scan the dataset for frequently occurring itemsets. Apriori is a commonly used algorithm, and it is quite efficient at finding rules in transaction data. Apriori uses a strategy called "level-wise" search. Beginning at the first level, we have individual items such as milk that occur frequently. At the next level, we find two-itemsets, such as bread and milk, where each of the component items meets the minimum support requirement. The algorithm keeps adding items to sets until there are no more combinations to be found that meet the minimum support. For example, let's say that we have Level 2 sets {Bread Tea}, {Bread Fish}, and {Tea Fish} with sufficient support. At Level 3 we might be able to join these together to produce the three-itemset {Bread Tea Fish}. In Python, this is the job that is accomplished by the apriori() function call. This function searches the transactions data for combinations of items that occur with sufficient frequency to meet the minimum support. These are often referred to by data miners as "frequent itemsets."

After obtaining a list of frequent itemsets, we then take a second step that creates the association rules in the form of "if antecedent, then consequent." So, each rule states that when the antecedent itemset occur(s), the consequent item also occurs a certain percentage of the time. To reiterate a definition provided earlier in the chapter, support for a rule refers to the frequency of cooccurrence of both members of the pair, that is, antecedent and consequent together. The confidence of a rule refers to the proportion of the time that antecedent and consequent occur together versus the total number of appearances of the antecedent itemset. For example, let's say that Milk is our antecedent, and it occurs in 25 percent of carts. Then we find out that Milk->Bread appears in 10 percent of carts. The confidence of the Milk->Bread rule would be 0.10/0.25 = 0.40. The following code uses these ideas to produce frequent itemsets and association rules for the Adult data:

```
te = TransactionEncoder()
te_ary = te.fit(adult_UCI_list).transform(adult_UCI_list)
df_adult_UCI = pd.DataFrame(te_ary, columns=te.columns_)
frequent_itemsets = apriori(df_adult_UCI,
              min_support=0.20, use_colnames=True)
rules = association_rules(frequent_itemsets,
```

```
                    metric="confidence", min_threshold=0.85)
print(rules.shape)
rules.head(5)

(100, 9)
```

It may seem weird that we go to all the trouble of running TransactionEncoder() to create our transaction list only to immediately convert that back to a pandas DataFrame, but a DataFrame is what the apriori() function expects, so that is what we need to provide. You will also notice that in the call to apriori(), we used a minimum support level of 0.20. Then in the call to association rules, we used a minimum threshold of 0.85 for confidence. These values were selected after examining frequency plots like the one shown earlier for the categories of education *and* after some experimentation to see how many rules were generated. In the one line of output shown in the previous code block, we see that we have generated 100 association rules. Tastes vary among data analysts, but 100 rules provide enough possibilities of some interesting rules without being so many that no human analyst would have the time to examine them. An excerpt of the output of the rules.head(5) call appears in Table 12.1.

In Table 12.1, the antecedents column is the if portion of each rule, whereas the consequents is the then portion of each rule. Support and confidence are as we have defined them. Lift is a measure of interestingness that we can use to screen the goodness of a rule. This table leaves out a few elements that will appear when you run the code in Python, most notably two other measures of interestingness called "leverage" and "conviction." All three of these measures—lift, leverage, and conviction—make comparisons between the observed level of support and what would be expected if the antecedent and consequent were not at all connected with each other. We don't need to dig into their formulas, but lift and conviction should be greater than one and the larger the better. Leverage is in the range of –1 to 1, and we would like the values to be as far from zero as possible.

There is one other issue to mention about the table above: For most of the work that data miners do with association rules, the consequent part of the equation contains one

TABLE 12.1	■ Support, Confidence, and Lift for Five Rules From the Adult Data Set			
antecedents	consequents	support	confidence	lift
(Female)	(<=50K)	0.295299	0.890749	1.170931
(Never_married)	(<=50K)	0.314975	0.95452	1.254762
(<=50K)	(United_States)	0.678473	0.891885	0.993828
(>50K)	(Married_civ_spouse)	0.204414	0.854283	1.864465
(>50K)	(United_States)	0.218951	0.915034	1.019622

item, like Bread. In contrast, the antecedent part can and will contain multiple items. A simple rule might have Milk in antecedent and Bread in consequent, but a more complex rule might have Milk and Bread together in antecedent with Tea in consequent. The first five rules show only one item in the antecedent here, but if you screen additional rules starting with rule 19, you will start to see pairs of items in the antecedent. Change the last line of code in the previous cell to "rules.head(25)" to see several rules with two items in the antecedent.

VISUALIZING AND SCREENING ASSOCIATION RULES

If you are using Python in a notebook, you may have the option of viewing an interactive table where you can sort the rows by a value in a column. If you have this capability, try sorting in descending order based on lift. The most interesting rules should bubble to the top of the list. Alternatively, we can plot the support, confidence, and lift of each rule. The following code makes that happen:

```
support = rules['support'].tolist()
confidence = rules['confidence'].tolist()
lift = rules['lift'].tolist()

plt.scatter(support, confidence, c=lift, alpha=0.5,
            marker="*", cmap="Blues_r")
plt.xlabel('support')
plt.ylabel('confidence')
plt.show()
```

This code produces the plot shown in Figure 12.4. Even though the resulting graphic shows a two-dimensional plot, we actually have three variables represented here. Support is on the X-axis, and confidence is on the Y-axis. All else being equal, we would like to have rules that have high support and high confidence. We know, however, that lift serves as a measure of interestingness, and we are also interested in the rules with the highest lift. On this plot, the lift is shown by gradations of grey shading. The darker grey the symbol, the closer the lift of that rule is to 1.86, which appears to be the highest lift value among these 100 rules.

The other thing we can see from Figure 12.4 is that although the support of rules ranges from somewhere below 2 percent up to 8 percent, the rules with high lift have support below 4 percent. Contrastingly, there are rules with high lift and high confidence, which sound quite promising. We can easily filter the rule set based on these values, and we can also use filtering to focus attention on particular consequents of interest:

FIGURE 12.4 ■ Visualization of Support, Confidence, and Lift for Adult Dataset

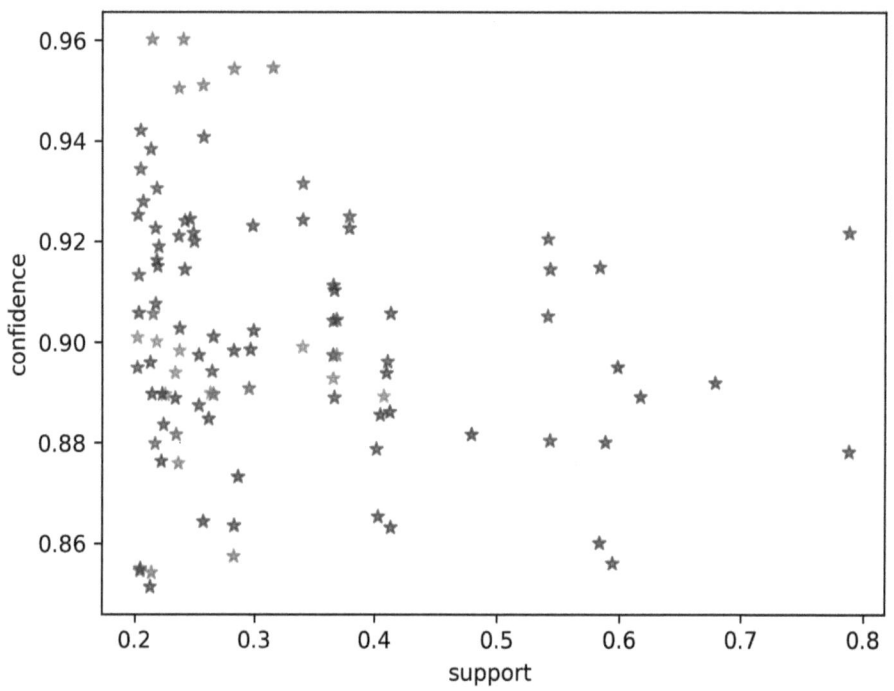

```
income_rules = \
            rules.loc[rules.consequents=={'<=50K'}].copy()
print(income_rules.sort_values(by='lift', \
                            ascending=False).head(1))

antecedents                                consequents
(Never_married, Private, United_States)    (<=50K)

support          confidence         lift        leverage     conviction
0.214119         0.960154           1.262168    0.044475     6.0052
```

In previous analyses of the Adult dataset, we were interested in demographic factors that predicted income status. If we filter the rules to a subset that includes Income<=50K, we can see the rule with the highest lift that remains in the filtered ruleset. A person who has never been married, is privately employed, and native to the United States has a high likelihood of having an income less than 50K in this dataset.

The key takeaway point here is that using a good visualization tool and filtering to examine the results of a data mining analysis can enhance the process of sorting through the evidence and making sense of it. If we were to present these results to a manager (and we would certainly do a little more digging before formulating our final conclusions), we would report the rules that had the highest lift values and that illustrated a pattern of results that was relevant to the insights we were seeking.

CASE STUDY: EXPLORING ASSOCIATION RULES WITHIN THE AIRLINE SURVEY

```
Case Key Points:
  - Transform our dataset to be able to apply rule mining
  - Create/filter association rules to examine detractors
```

As with the AdultUCI dataset, to create association rules from our air passenger survey dataset, we need to create one that has only categorical attributes and then convert it to a transactions dataset. Some attributes, such as Airline.Status, are easy to make into labeled attributes, but others, such as Age, require some feature engineering. For example, we likely want to use the same label for people in their early 30s but a different label for people in their 70s. One approach for handling Age is to round the attribute. This would mean for Age, we are looking for rules based on a person's decade. This approach can also work for other similar attributes such as Shopping.Amount.at.Airport. Here is code to create attributes from our survey, to which we then can apply rule mining:

```
survey = survey.loc[:,['Detractor','Age',
                       'Price.Sensitivity',
                       'Shopping.Amount.at.Airport',
                       'Class', 'Type.of.Travel',
                       'Airline.Status', 'Gender',
                       'Origin.State']]

survey = survey.rename(columns={'Price.Sensitivity':
                       'Price_Sensitivity',
                       'Shopping.Amount.at.Airport':
                       'Shopping_Amount_at_Airport',
                       'Type.of.Travel': 'Type_of_Travel',
                       'Airline.Status': 'Airline_Status',
```

```
                   'Origin.State': 'Origin_State'})

survey['Price_Sensitivity'] = ["Price_Sensititivity=" +
            str(x) for x in survey['Price_Sensitivity'] ]

survey['Detractor'] = ["Detractor=" +
            str(x) for x in survey['Detractor'] ]

survey['Shopped_at_airport'] = "no_shopping"

survey.loc[survey['Shopping_Amount_at_Airport'] > 0,
            'Shopped_at_airport'] = "shopping"

survey=survey.drop(columns=['Shopping_Amount_at_Airport'])

survey['Age'] = round(survey['Age'],-1)

counts = survey['Type_of_Travel'].value_counts()
most_occuring = counts[counts>1000]
most_occuring.plot(kind='bar')
fig = ax.get_figure()
plt.savefig('Figure12_5.pdf',dpi=300,bbox_inches='tight')
```

This code accomplishes several transformations. First, we subset the columns to nine variables of particular interest. Next, we rename the columns to change the dots to underscores: This avoids confusing the TransactionEncoder(). Then we encode brief strings for Price_Sensitivity, Detractor, and Shopped_at_airport. These changes provide unique labels for each attribute of interest. Then we round the Age variable to the nearest decade (i.e., 20, 30, 40, etc.). Finally, we drop the numeric Shopping_Amount_at_Airport because we have recoded that into a binary variable Shopped_at_airport. The last three lines of code in the previous block produce a diagnostic graph for Type_of_Travel so that we can check out the relative frequency of each of its categories. We can see the result in Figure 12.5. You should produce this kind of diagnostic plot for every column you plan to use with TransactionEncoder():

This graph shows that more than 50,000 of our rows pertain to business travel, with personal travel coming in second and miles or reward travel a distant third. With more than 5,000 instances in that last category, however, we may still get some interesting rules that include it. When you examine a graph like this, you can get a sense of the minimum support values you might need to include the attributes. If we set minimum support much higher than about 6 percent in this case, we won't see any rules that involve mileage travel. In the next code block, we use 6 percent as our minimum threshold for support, and we end up with 764 frequent itemsets. Note that the following code block follows three chunks of code

FIGURE 12.5 ■ **Diagnostic Plot of Travel Type for Air Passenger Survey Data**

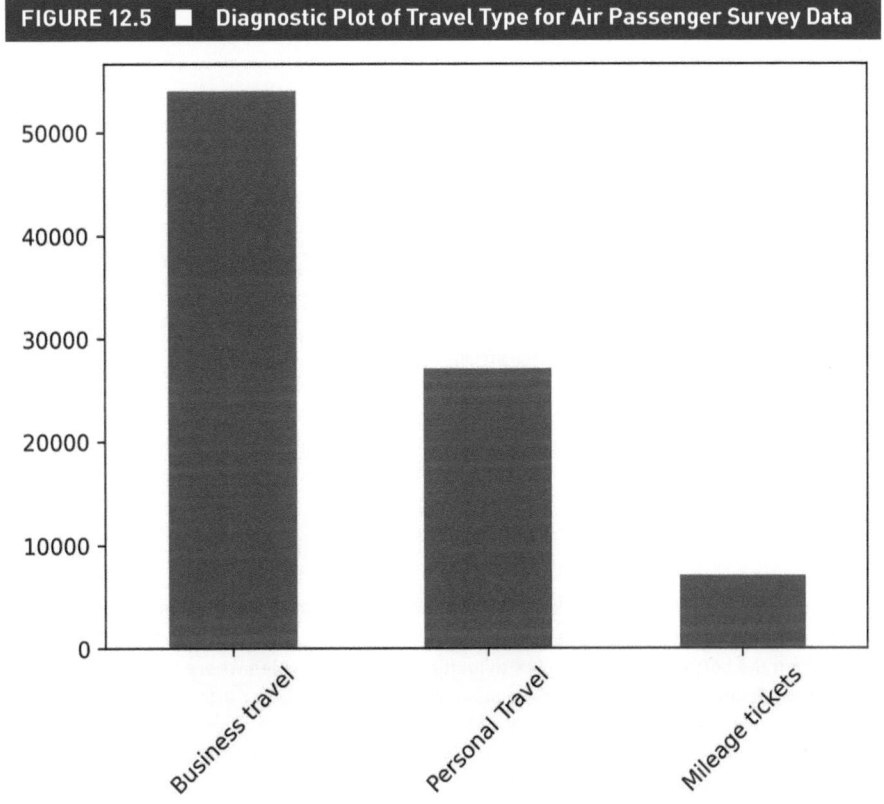

that were demonstrated earlier in the chapter: substituting underscores for dashes in the attribute names, recoding each row of DataFrame data into a list of attributes, and running TransactionsEncoder(). Consult the notebook for this chapter to examine the code used in those steps.

```
frequent_itemsets = apriori(df_transactions_survey,
                    min_support=0.06, use_colnames=True)
print(len(frequent_itemsets))
rules = association_rules(frequent_itemsets,
                    metric="confidence", min_threshold=0.3)
rules = rules[rules['consequents'] == {'Detractor=True'}]

print(rules.shape)
rules.head()
```

```
764
(63, 9)
```

TABLE 12.2 ■ Excerpt of Rules Where the Consequent Is Detractor=True

antecedents	consequents	support	confidence	lift
(Blue)	(Detractor=True)	0.253882	0.371837	1.259414
(Female)	(Detractor=True)	0.182312	0.322912	1.093705
(Personal Travel)	(Detractor=True)	0.193062	0.627092	2.123965
(Price_Sensititivity=2)	(Detractor=True)	0.095736	0.344484	1.166769
(no_shopping)	(Detractor=True)	0.180451	0.315887	1.06991

The output in this code block confirms that we generated 764 frequent itemsets and that a minimum confidence threshold of 0.3 generates 69 rules. Note that the third line of code from the bottom filters the rules to focus only on those rules where the consequent is Detractor=True. This filtering provides an easy way of focusing our attention on issues that may adversely affect customers. Table 12.2 excerpts the first five rules that pass this filter.

The consequents (consequent) are all the same because of the filtering we applied. All of these rules have lift greater than one. We can see some of the attributes associated with a customer being a Detractor, such as price sensitivity and no shopping at the airport. When we report on findings like this, it is important to remember that we cannot make any causal statements. It is sensible that a price-sensitive customer would not shop at the airport because prices are higher there, but we don't know that either of these attributes cause a person to be a Detractor. The connection among these attributes is likely to be much more complex. To close out this section, let's make a support/confidence and lift graph:

```
support = rules['support'].tolist()
confidence = rules['confidence'].tolist()
lift = rules['lift'].tolist()

plt.scatter(support, confidence, c=lift, alpha=0.5,
            marker="*", cmap='Blues')
plt.xlabel('support')
plt.ylabel('confidence')
plt.savefig('Figure12_6.pdf',dpi=300,bbox_inches='tight')
plt.show()
```

This code produces the graph shown in Figure 12.6. This graph shows distinctive separated patterns of high confidence and low confidence rules.

FIGURE 12.6 ■ Support, Confidence, and Lift for Association Rules Where Detractor=True

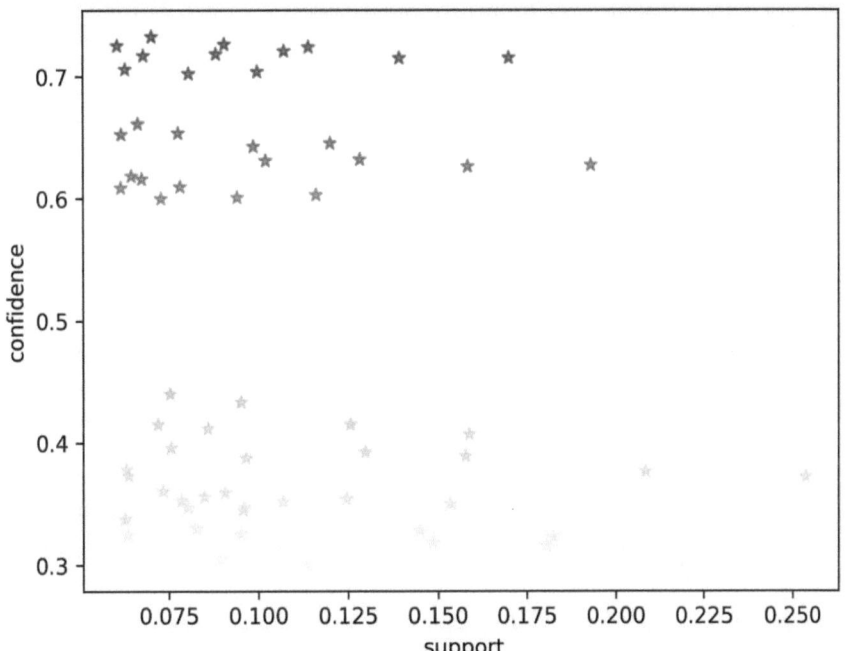

It may have struck you that by setting our consequent to be Detractor=True, we have conducted an unsupervised data mining analysis that could highlight some promising predictors for a supervised analysis. This line of code displays the antecedent of the rule with the greatest lift:

```
rules.sort_values(by='lift',\
                  ascending=False).antecedents.head(1)

(Blue, Female, no_shopping, PersonalTravel)
```

The antecedent of this high lift rule includes the frequent flier status (Blue), sex (female), shopping behavior at the airport (no_shopping), and type of travel (Personal_Travel). These predictors could be explored later in a supervised analysis. This simple example shows that unsupervised and supervised data mining techniques are not mutually exclusive but rather can be complementary to one another.

CHAPTER CHALLENGES

1. The GitHub repository contains the full Groceries dataset that was introduced early in the chapter but not fully analyzed. Prepare the data for analysis and examine an association rules analysis. For frequent itemsets with the minimum support set to 0.1, what is the rule with the highest confidence?

2. Continue your analysis of groceries by revealing which association rule or rules have the highest lift value.

3. Filter the association rules from exercise 1 to focus only on rules with a minimum confidence of 0.5. How many such rules are there?

4. Plot the rule set from the previous problem. Make sure to show support on the X axis, confidence on the Y axis, and shading to show the level of lift. Describe the pattern you see.

5. Conduct a new analysis of the Groceries dataset, this time forcing the consequent to be {Milk} using the minimum support of 0.1 and minimum confidence of 0.75. Report how many rules were generated, create a plot of the rules, and report the maximum lift value.

6. Filter the rule set created in the previous problem to focus on the highest lift rules. Choose a lift value for filtering based on the maximum lift value such that fewer than 10 rules are reported.

7. Based on the smaller high lift rule set generated in the previous problem, report the most complex rule (i.e., the one with the most items in the antecedent). Make sure to report the support, confidence, and lift for that rule. Describe in your own words what that rule is saying and why it makes sense.

8. Conduct a complete association rules analysis of the air passenger data using the code from the online notebook to get started. Instead of focusing on Detractor=True, filter your rules with Detractor=False, and report the highest lift rules that you obtain.

13 WORDS OF WISDOM: DOING TEXT ANALYSIS

LEARNING OBJECTIVES

Access and analyze unstructured data.

Demonstrate accessing and analyzing unstructured data using word clouds.

Apply Python libraries to conduct basic text mining.

Code and use the sentiment analysis text mining technique.

Code and use a topic modeling technique.

UNSTRUCTURED DATA

Prior chapters focused on analysis of highly structured datasets. This included numeric and categorical data types. With highly structured datasets, we usually have a set of variables as columns, and a set of observations as rows, where each row contains values in most or all of the columns. Our diagnostics and visualizations focused on understanding each variable to facilitate analysis and interpretation.

This chapter switches gears to focus on manipulating datasets that do not offer such convenient, well-ordered data structures. The three main types of data that fit in this category are images, video, and text. In this chapter we will focus only on that third type. Text data are everywhere and can include books, news articles, speeches, discussion boards, and web pages. Additional sources of text include social media platforms as well as product and service review platforms. Companies often provide surveys to their customers, and sometimes those surveys include text boxes where customers can type a more detailed response. These "natural language" texts can provide a gold mine of useful information for understanding people. Finally, there has been a proliferation of dialog engines, also known as "chatbots," that conduct text-based conversations with people, and the texts generated by these interactions may also contain useful information.

So, let's begin our exploration of natural language text data. The picture at the start of this chapter shows an example of a diagnostic visualization of a text dataset. The visualization is a word cloud using a famous speech as input. Word clouds are fun to look at, but they also contain useful diagnostic information. The geometric arrangement of words on the figure is partly random and partly designed and organized to please the eye. The font size of each word conveys some measure of its importance in the corpus that was presented to the word cloud plotting code. The word "corpus," from the Latin word for body, is used by text analysts to refer to a body of textual material, usually consisting of multiple documents. When thinking about a corpus of textual data, a set of documents could be anything: web pages, word processing documents on your computer, a set of tweets, or a set of government reports. In most cases, text analysts examine a corpus where all of the documents within fit into the same genre. An example of a corpus of interest to a company would be a collection of 1,000 written customer reviews of one of their products. It would be rare to mix different genres of text in the same corpus—for example, we would not typically want to mix customer reviews and tax disclosure documents in the same corpus.

Similar to visualizations like histograms and scatterplots, word clouds are useful when trying to get a quick understanding of some text data. When we want a quick overview of a speech or essay, a word cloud highlights the most frequently occurring words. In this chapter, we will use some Python code to investigate the text from a speech and create a word cloud like the one shown at the start of this chapter. Before beginning our work with Python libraries, we need to read in a text source. As an example, we will explore a famous historical speech by Susan B. Anthony. Prior to 1920, women in the United States lacked the right to vote in federal elections. Famous suffragist Susan B. Anthony spoke publicly after being arrested for voting in the presidential election of 1872, an unlawful act at the time. You can read the speech at http://www.historyplace.com/speeches/anthony.htm. We can use Python to read the text directly from this web page.

READING IN TEXT FILES

To get started with our new project, the first task we need to do is import the text we want to analyze. As you have learned, there are many different ways to read data into Python. Until now, we have used techniques for reading in CSV files and a few other formats that are good at storing structured data. If we want to read in unstructured (text-based) data, we need some new functions, especially if we are extracting text from a web page.

As shown in the next cell, we can read directly from a web page into Python string. If you have ever examined a web page in its most raw form, you know that it is full of special labels, called "tags," that make it possible for your browser to display and format the page properly. These tags are part of a computer language called hypertext markup language (HTML). We must process the HTML and convert it into simple text by removing the tags. To do this, we can use the urllib library to read a web page and the BeautifulSoup library to strip out the HTML tags.

```
from urllib.request import urlopen

url = 'http://www.historyplace.com/speeches/anthony.htm'
html = urlopen(url).read()
html


b'<!doctype html>\n<html>\n<head>\n<meta charset="utf-8">\n


from bs4 import BeautifulSoup

soup = BeautifulSoup(html, features="html.parser")

for script in soup(["script", "style"]):
    script.extract()

text = soup.body.get_text(separator='\n', strip=True)

text


'In the 1800s, women in the United States had few legal rights and
did\n not have the right to vote. This speech was given by Susan B.
Anthony after\n her arrest for casting an illegal vote in the presiden-
tial election of\n 1872. She was tried and then fined $100 but refused
to pay.\nFriends and fellow citizens: I stand before you tonight under
indictment\n for the alleged crime of having voted at the last presiden-
tial election. . .'
```

The first chunk of code uses the read() method of urlopen to extract the text from the specified web page. The second chunk of code instantiates a BeautifulSoup object and initializes it with a copy of the HTML that we read from the web page. We then have a for loop that uses the extract() function to remove the coded elements of the web page that we do not need. Finally, we use the get_test() method to extract the text from the main body of the web page. Note that our display of the text at the end of that code cell has been truncated: You will probably see more of the speech in your Python environment although not necessarily the whole speech. You can readily see, however, that all of the HTML tags have been stripped out, leaving the plain text of the speech plus some unwanted extra spaces and newline characters. We can easily remove those as well, by using a list of the punctuation to be removed, which can be imported from the string library:

```
import string

for s in string.punctuation:
    text = text.replace(s, '')

words = text.split()

words = [word.lower() for word in words]
words = words[:-149] # Trim out the bottom of the web page
words[:10]

['in', 'the', '1800s', 'women', 'in', 'the', 'united', 'states', 'had', 'few']
```

In these steps, we have imported the string package, used a for loop to remove punctuation marks, divided the text by spaces into individual words, and made everything lowercase. That last step is a process that text analysts call "normalization," and it is used intentionally to make each word consistent. The result is a list, in which each element contains a string. You can see when examining the first 10 elements of the list that each element is a string containing letters and numbers. Analysts often refer to these elements as "tokens."

CREATING THE WORD CLOUD

Remember that text mining focuses on extracting useful analytic information from corpora of text (the word "corpora" is the plural of corpus). Although some people use the terms "text mining" and "natural language processing" interchangeably, there are differences worth remembering. First, the "mining" part of text mining refers to an area of analytic practice that looks for patterns in large datasets, or what some people refer to as "knowledge discovery" in databases. In contrast, natural language processing reflects an interest in understanding how machines can be programmed (or can learn on their own) how to understand and/or produce human language. Thinking about the difference from another angle, natural language processing tends to maintain and use linguistic information and therefore to organize text into its component grammatical elements such as nouns, verbs, adjectives, and adverbs. Text mining usually focuses on statistical approaches to analyzing text, using strategies that involve counting the frequency of occurrence of tokens regardless of the part of speech they represent or the order in which they appeared in a sentence. A word cloud is a graphical presentation of word frequencies and is therefore helpful as a starting point in a text mining analysis.

Before we create our word cloud, we should filter our stream of tokens. Researchers who developed the early search engines for electronic databases found that words such as "the," "a," and "at" appeared so commonly in so many different parts of the text that they interfered with the creation of a search engine. In contrast, the unique nouns, verbs, and adjectives that appeared in a document did a much better job of setting a document apart from other documents in a corpus, so researchers decided that they should filter out all of the short, commonly used words. The term "stop words" seems to have originated in the 1960s to signify words that a computer processing system would stop using because they had little meaning in a data processing task. In most languages stop words don't carry much meaning, although they obviously have important roles in speech and writing.

If we fail to remove stop words from our stream of tokens, our word cloud will not be helpful. As an example, "the" is usually the most frequently occurring word in any corpus of English text. If our word cloud had a giant THE in the middle, it would not help us understand the Susan B. Anthony speech. As preparation for making the word cloud, let's filter out the stop words:

```
from collections import Counter

Counter(words).most_common(4)

[('the', 54), ('of', 32), ('to', 25), ('and', 21)]

import nltk
from nltk.corpus import stopwords
# You might also have to download the stopwords
nltk.download('stopwords')

stop_words = stopwords.words('english')
words = [w for w in words if not w in stop_words]

# look at top 5 most freq terms
Counter(words).most_common(4)

[('women', 8), ('states', 6), ('citizens', 6), ('united', 5)]
```

In the first step here, we use the Counter class to examine the most frequent words. As we predicted, stop words like "the" are at the top of the list. Then we download and use a list of stop words from the Natural Language Toolkit (NLTK) to filter all of the stop words from the token stream. NLTK is a workhorse among Python libraries, and it contains hundreds of useful resources that text analysts frequently use. In the final line of code, the second call to Counter() confirms that we have eliminated stop words, leaving the more relevant tokens such as "women" and "states" as the most commonly occurring ones.

Now we are ready to generate our word cloud. We can use the WordCloud library for this purpose:

```
from wordcloud import WordCloud

wordcloud=WordCloud(colormap="Blues",\
                    background_color="white",max_words=40)

sentence = ' '.join(words)

wordcloud.generate(sentence)

wordcloud.to_image()
```

After importing the WordCloud package, we instantiate a WordCloud object, providing arguments to specify the colors and the maximum number of words. When we call on this object to generate the word cloud, it expects a single string of text rather than a list of tokens, so the next line of code rejoins the tokens into one long string with a space between each word. Finally we process the text using the generate() method on the wordcloud object and show the resulting plot with the to_image() method to produce the graphic shown in Figure 13.1.

Word clouds represent an important first step when starting to examine a corpus of text data. In more complex datasets, a word cloud will help root out nuisance words, such as misspellings or acronyms, that you may want to filter out prior to further analysis. This basic word cloud looks fine, so we will now take the next analytic step by using another technique for textual data known as "sentiment analysis."

SENTIMENT ANALYSIS

Sentiment analysis sounds complex, and in fact, there are many ways to do sentiment analysis, but we will use a simple strategy based on finding and counting keywords. Our process will use dictionaries of positive and negative words in a library known as Valence Aware Dictionary and sEntiment Reasoner (VADER). The researchers who developed VADER focused their attention on analyzing short fragments of text, such as those found in social

FIGURE 13.1 ■ **Word Cloud of Susan B. Anthony Speech**

media postings. The VADER software was later integrated into NLTK, making it convenient for us to access.

As a preprocessing step before conducting our sentiment analysis, we can divide our Susan B. Anthony speech into sentences and pretend that she sent each one as a separate social media posting. When you are analyzing text, it is up to you to decide how big a chunk represents a single unit. Common units of analysis include sentences, paragraphs, posts, chapters, or complete documents. The choice depends mainly on the goal of an analysis: If we analyze a book paragraph by paragraph, we will learn something different then if we analyze it chapter by chapter. We will use the NLTK sentence tokenizer, which in turn depends on a word-by-word tokenizer also from the NLTK library, known as "punkt." Punkt is the German word for the period or dot character, and the authors of the original paper that led to the development of the punkt code were from a university in Bochum, Germany.

```
nltk.download('punkt')

text = soup.body.get_text(separator='\n', strip=True)

text = text.replace('\n', ' ')

sentlist = nltk.sent_tokenize(text)

# Remove the end matter of the web page
```

```
    sentlist = sentlist[:-3]

    print(sentlist[2])
    print(sentlist[5])

 She was tried and then fined $100 but refused to pay.
    The preamble of the Federal Constitution says: "We, the people of the
 United States, in order to form a more perfect union, establish jus-
 tice. . .
```

Let's review what we've done. After downloading the punkt tokenizer, we returned to our "soup" object that contains the web page and used it to recreate our large text string. We had to do this because our earlier preparations for the word cloud had stripped out punctuation from the text string, leaving no signals to help the sentence tokenizer divide the text into sentences. Next, we filtered out the newline characters, because these can sometimes confuse the sentence tokenizer. Then we called the NLTK sentence tokenizer, which returned a list of text strings, where each string contained a single sentence. The web page had a bunch of stuff in the footer that didn't belong in the speech, so we eliminated the last three sentences from the list. Finally, we showed two sentences from the list. At this point we were ready to instantiate a VADER object and analyze the sentiment of each of our sentences:

```
from nltk.sentiment.vader import SentimentIntensityAnalyzer

nltk.download('vader_lexicon')

sia = SentimentIntensityAnalyzer()

sentiment_list = []  # Start an empty list

for sent in sentlist:
    sentiment = sia.polarity_scores(sent)
    sentiment_list.append(sentiment['compound'])

sentiment_list[2], sentiment_list[5]

(-0.5267,0.9802)
```

The second line of code here downloads the lexicon that VADER needs to do its job. A "lexicon" is a specialized dictionary that contains extra information useful for some kind of analysis. In this case the VADER lexicon contains a dictionary of common words and their sentiment scores. Each sentiment score in the lexicon was developed by a team of human raters who agreed on a score ranging between −4 and +4. In the next line of code, we instantiate the sentiment analyzer. Then we use the analyzer in a for loop to process every sentence in our list. Each time we call the polarity_scores() method, we get back a percentage of positive, neutral, and negative features in the sentence along with an overall sentiment score known as "compound." The compound score takes into account all of the positive and negative features of a sentence and then computes a result that is bounded between −1 and +1. You can see from the results at the end of the code block that we have reported the compound score for one negative sentence and one positive sentence. These happen to be the same two sentences whose text is shown in the previous cell block. The negative sentence contains "tried and fined but refused to pay," a heroic move on Susan B. Anthony's par, but linguistically a kind of negative scenario. In contrast, the first sentence of the preamble of the constitution contains "perfect union," "establish justice," and several other lofty phrases that VADER scores as containing strongly positive sentiment.

One important thing to remember about dictionary-based sentiment analysis is that the nature and quality of the results are highly dependent on how the dictionary was developed. We would not want to use VADER to analyze quarterly earnings reports because its lexicon was designed for use on short passages of text, and it does not contain a wide variety of financial terminology. Fortunately, there are other sentiment methods to explore, such as SentiWordNet scoring (also available in NLTK) that may be applicable when VADER is not. We recommend that you try more than one method of sentiment analysis on your text and compare the results to see which system provides the most meaningful outputs.

It is also possible to use training data to develop your own sentiment scoring system. To begin this process, you would need a database of text examples, where each example had already been scored by a human rater. Numerous Python coding examples are available online to demonstrate how this can be accomplished. Some sentiment system, like VADER, also have methods by which you can add your own custom words to the existing sentiment dictionary.

TOPIC MODELING

The sentiment analysis we did was quick and easy: We processed each sentence from the Susan B. Anthony speech using calls to the VADER sentiment scoring system built into NLTK. Although simple, this approach can help us draw some general conclusions about the overall tone of a social media postings, short reviews, or other brief documents. To dig a little deeper, though, we need to use statistical techniques that reveal more details about the individual documents in our corpus.

One family of such techniques is called "topic modeling." Topic modeling creates a set of statistical profiles of word frequencies: Each profile in the set is called a "topic," and every document in the corpus is represented as a mixture of these topics. Like the dictionary-based sentiment analysis we did, topic modeling uses the frequencies of words to make sense out of a corpus of documents. Unlike sentiment analysis, topic modeling allows us to create unique profiles of word frequencies.

To make the statistical processing work properly, every topic actually contains at least a small contribution from every word in the corpus. In a good topic model, however, certain words make each topic distinctive by contributing strongly to that topic and that topic alone. Every document is considered to be a mixture of all the topics, but in a good model each document is represented strongly by only one topic. Consider the first two lines of the Robert Frost poem "In a Poem" shown in the center of Figure 13.2. If we treat each line of the poem as a document, we have a corpus with two documents and eight features (the features are the bolded words in each line of the poem—stop words have been removed). If we create a topic model containing two topics, topic 1 contains the terms "sentencing," "goes," "blithely," and "way," with 25 percent weight for each. Topic 1 also contains the words "takes," "playfully," "objected," and "rhyme," but these all have a weight of zero. Topic 2 completely reverses that system of weights.

Given these two distinctive profiles of words, we can say that document 1 (the first line of the poem) is composed of 100 percent topic 1 and 0 percent topic 2. Document 2 is composed of 0 percent topic 1 and 100 percent topic 2. Of course, a more complex set of documents and terms won't ever turn out this simply. Unlike this example, which has two documents and two topics, we're also hoping to represent a set of documents using a relatively small number of topics—for example, 30 topics to represent 1,000 restaurant reviews. You could imagine using this method with product reviews, or web pages, or social media postings to automatically sort these kinds of documents into groups based on whether the customer has a complaint, compliment, suggestion, or something else. Let's apply topic modeling to the sentences of the Susan B. Anthony speech and see what we get:

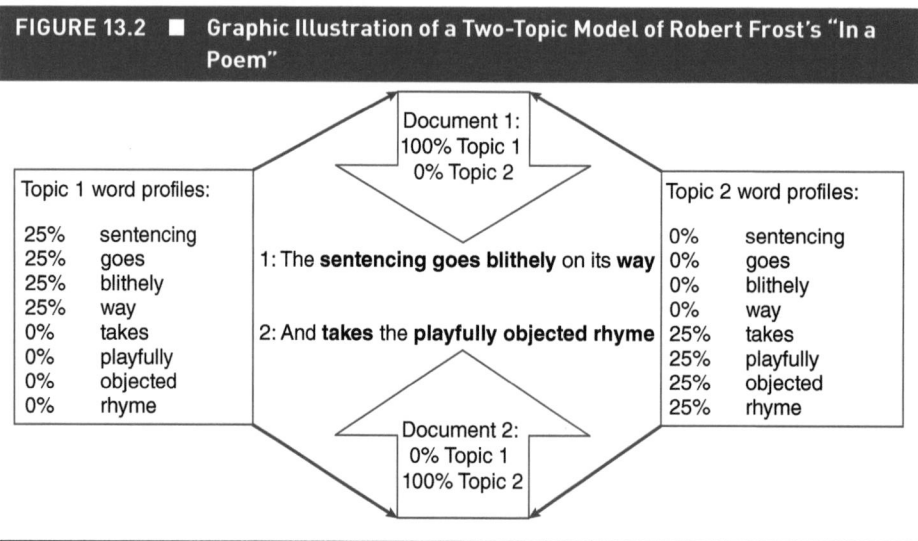

FIGURE 13.2 ■ Graphic Illustration of a Two-Topic Model of Robert Frost's "In a Poem"

Note. From *Data Science for Business With R,* by J. S. Saltz and J. M. Stanton, p. 315. Copyright 2022 by Sage Publications. Reproduced with permission.

```
from sklearn.feature_extraction.text import CountVectorizer
from nltk.tokenize.casual import casual_tokenize
import string

for i, sent in enumerate(sentlist):
    sentlist[i] = "".join([ch for ch in sent \
                         if ch not in string.punctuation])

counter = CountVectorizer(min_df=2, max_df=0.9,
                          tokenizer=casual_tokenize,
                          stop_words='english')

countmatrix = counter.fit_transform(raw_documents=sentlist)

countmatrix.shape, type(countmatrix)

((21, 29), scipy.sparse.csr.csr_matrix)
```

In the first for loop, we remove punctuation characters—otherwise they will show up in the topic model. Then, we instantiate a CountVectorizer, which is a class for creating a data structure known as a document-term matrix. In creating the instance of this class, we create a filter specifying that a term must appear in a minimum of two documents and no more than 90 percent of all documents to be included in the data structure. This strategy protects against super rare words that appear only once and common words that will tend to appear in all documents. In this step we also remove all stop words such as "the." Finally, we send our list of sentences into our vectorizer, using the fit_transform() method, to produce the document-term matrix. The diagnostics at the end show that we have 21 sentences and 29 vocabulary words (terms) in the matrix. Each row of the matrix represents one sentence, and each column represents one term. The cells of the matrix contain the counts of how frequently each term appears in each sentence.

Next, let's run an algorithm called Latent Dirichlet Allocation (LDA). This algorithm uses a statistical model to create the topics and assign the topic-document probabilities and topic-word probabilities. The "Dirichlet" in the title refers to a statistical distribution that is good for modeling probabilities. Like many other modeling techniques, we have to specify the number of topics that we want in advance. Before running the analysis, however, we don't know what number of topics might produce a good result. In the code that follows, we run three different models, with four, five, and six topics respectively, and then use a measure called "perplexity" to choose among them. We want perplexity to be as small as possible. Choosing the range of four to six topics was a guess from experience: You can easily modify the code to screen a larger set of possibilities if you are unsure of the appropriate range to test. Also keep in mind that there is no perfect number of topics for any given dataset. We want to achieve a balance between interpretability, which is often easier with a smaller number of topics, and fit, which typically improves as we add more topics.

```
import pandas as pd
from sklearn.decomposition import LatentDirichletAllocation

countdf = pd.DataFrame.sparse.from_spmatrix(countmatrix,
                columns=counter.get_feature_names_out())

lda4 = LatentDirichletAllocation(n_components=4, \
                                 random_state=1)
lda4.fit_transform(countdf)

lda5 = LatentDirichletAllocation(n_components=5, \
                                 random_state=1)
lda5.fit_transform(countdf)

lda6 = LatentDirichletAllocation(n_components=6, \
                                 random_state=1)
lda6.fit_transform(countdf)

lda4.perplexity(countdf) \
lda5.perplexity(countdf) \
lda6.perplexity(countdf)

(53.27855793562065, 46.55818564847087, 59.758787095278464)
```

We begin by turning our sparse matrix into a pandas DataFrame, which is what the LDA algorithm expects to receive. Then we instantiate and run the LDA with four, five, and six topics. Note the use of the argument "random_state=1." This controls the sequencing of random numbers to make the output repeatable. Otherwise, you will get a slightly different result every time because of the use of random numbers in initializing the LDA algorithm. The fit_transform() method does the work of fitting the topic model to the data. At the end of the code cell, we report the perplexity values generated by each model. The slightly smaller perplexity score for five topics suggests that this may be a reasonable choice.

The topic modeling process produces two large matrices of coefficients, called "beta" and "gamma," that together define how words, topics, and documents are connected with one other. The beta matrix shows the connection between words and topics and can be revealed with the components_ attribute on the fitted LDA object (e.g., lda5.components_). The gamma matrix shows the connection between topics and documents and can be computed using the transform() method and supplying the document term matrix as input, for example, lda5.transform(countdf). Although expert analysts may like to examine these numeric values in detail, we will take a more high-level approach and examine some topic visualizations produced by a Python library called "pyLDAvis."

```
import pyLDAvis
import pyLDAvis.lda_model
pyLDAvis.enable_notebook()

pyLDAvis.lda_model.prepare(lda5, countmatrix, counter)
```

Depending upon your notebook environment, you may need to install the pyLDAvis package before trying to import it. The third line of code here also performs some important setup chores so that the resulting graphic will display properly in a notebook environment. Finally, the last line of the cell proves three ingredients that pyLDAvis needs to create the visualization shown in Figure 13.3: the output object from the LDA analysis (which contains the beta and gamma matrices), the original document-term matrix that was used as input to the LDA process, and the vectorizer function used to create that document-term matrix.

Figure 13.3 shows a snapshot of the interactive display produced by pyLDAvis. This snapshot depicts the separation and importance of four of the five topics in the left-hand pane and a breakdown of predicted word frequencies for topic 4 in the right-hand pane. On the left, the larger the bubble, the more that topic appears across documents. The further away a bubble is from other bubbles, the more distinctive that topic is with respect to its profile of expected word frequencies. Note that the numbering of the topics in the graph does not match the

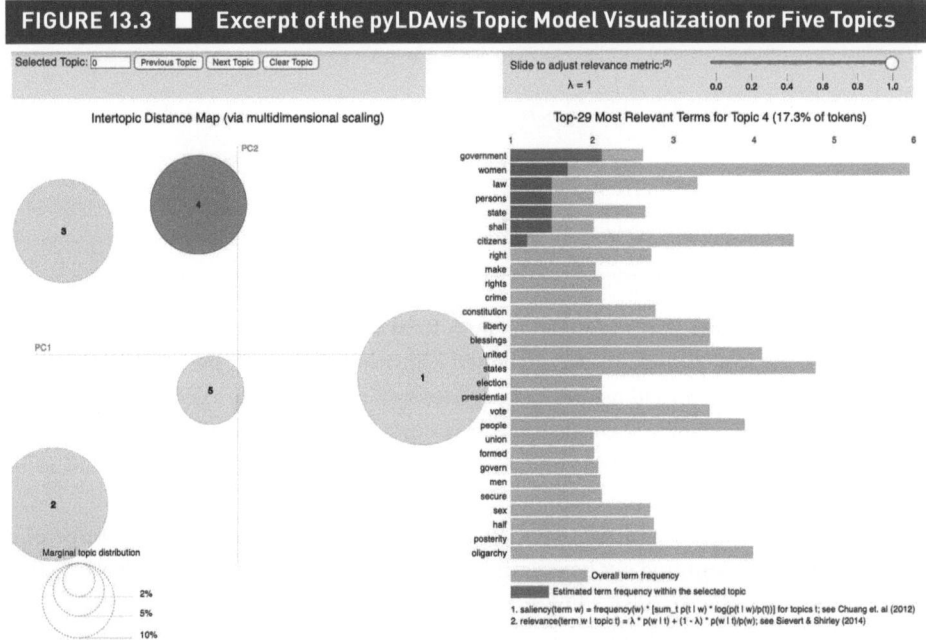

FIGURE 13.3 ■ Excerpt of the pyLDAvis Topic Model Visualization for Five Topics

numbering of topics in the computed analysis because the package was originally created for R, where indexing begins with 1 instead of 0. When you create the visualization in a notebook, you can use the buttons and slider at the top of the graph to change the display.

In the right-hand pane, the darker part of the bar for the topmost seven words shows their prevalence in the present topic, whereas the lighter part of each bar shows the overall prevalence of the word. In this case, the distinctiveness of topic 4 comes from the combination of government, women, law, persons, state, shall, and citizens. Keeping in mind that our analysis treated each sentence of the speech as a document, a sentence like this one at the conclusion of the speech has a strong connection to topic 4: "Being persons, then, women are citizens; and no State has a right to make any law, or to enforce any old law, that shall abridge their privileges or immunities."

Interpreting the results of the LDA analysis requires a lot of human judgment to make sense of the list of words that comprises a topic. You may need to provide that kind of human interpretation to data science clients, but most text analysts don't stop there. If you think about what the gamma matrix contains, each document in our list has a unique fingerprint based on the specific values it contains for each topic. In our example, each of the 21 sentences has a vector representation consisting of five floating point numbers (one for each topic). When two documents (in this case two sentences) are similar to one another with respect to the words they contain, their document-topic vectors will also be similar. When two documents are different, their vectors will also be different.

We can use those document-topic vectors (i.e., the contents of the gamma matrix) in subsequent analyses. Each column of the gamma matrix may contain information of interest that we can use for other purposes. Have a look at this code, correlating sentiment scores from earlier in the chapter with the document-topic vector for topic 0:

```
import numpy as np

doc_topic = lda5.transform(countdf)
np.corrcoef(sentiment_list, [d[0] for d in doc_topic])

array([[1.        , 0.66593129],
       [0.66593129, 1.        ]])
```

This output shows that the topic 0 values from the gamma matrix correlate at nearly r=0.67 with the sentiment scores we computed earlier in the chapter. By capturing common linguistic patterns in word usage, we have apparently tapped into more general aspects of the meaning conveyed by some of the documents in our corpus. Each column of the topic-document matrix might have something interesting to tell us about sentiment or a related issue like emotional content or particular customer service problems.

Next, because each set of five numbers in our example represents a unique fingerprint for a document, we can find out how similar two documents are quite easily. A basic method for establishing similarity is to compute the "dot product" between two vectors—the sum of the products of the individual elements. The Python statement "np.dot(doc_topic[19], doc_topic[17])" reveals the dot product between the vectors for document 19 and document 17. The resulting value of 0.68 suggests that these two sentences from the Susan B Anthony speech are quite similar. In fact, sentence 17 provides the rhetorical setup for the statement that Anthony makes in sentence 19: "The only question left to be settled now is Are women persons?" This method of assessing document similarity could be helpful in grouping a set of documents together for further analysis, indexing, or searching.

OTHER USES OF TEXT MINING

Although we have focused on using text mining for sentiment analysis and topic modeling, there are many other applications of text mining. Let's explore scenarios where text mining might be useful. The first common example is to do text mining on social media posts. By analyzing posts, we can explore the frequency of a certain hashtag or how and when a specific hashtag starts to trend. In addition, a company could use text mining to review social media posts to get an understanding of how consumers view their product.

In a different example, text mining could be used in call centers. For example, for an inbound call center (where people call in with questions or to order a product), it is helpful to know what people are talking about on the phone. Using speech-to-text, an organization can generate transcriptions of the phone conversations, which can then be analyzed using text mining to find keywords or topics representing types of calls. Armed with these data, the organization can then do analysis of what customers talk about during calls, potentially connecting the topics to needed product improvements. In a related use of text mining, notes about repairs (e.g., by a manufacturer) could be analyzed to determine if specific keywords (representing specific components in a manufactured system) are connected with product reliability problems.

To recap, in this chapter we explored sentiment analysis, a text mining technique that attempts to determine how positive or negative an unstructured text document is. We also reviewed a simple example of topic modeling, where we used a statistical analysis to create topics with unique word profiles and mapped those onto the documents in the corpus. We touched upon some other possible scenarios where text mining could be used, such as analyzing customer comments.

CASE STUDY: CONNECTING TOPICS TO NPS

```
Case Key Points:
 - The customer survey contains many brief text comments
 - Customer comments may contain key words that are
   associated with likelihood to recommend
 - We can create a topic model and then analyze the gamma
   matrix to predict likelihood to recommend
```

This chapter demonstrated how to use a corpus of documents to create topic models—a set of statistical profiles of word frequencies. In the case study dataset, a subset of customer flight segments contains brief comments that customers wrote on their surveys. In the analysis that follows, we will generate a topic model to ascertain whether any topics will correlate with the variable Likelihood.to.recommend. If there is a topic that does predict this outcome variable, we can examine examples of documents that represent this topic to obtain actionable insight about customer retention. First, let's read in the survey.

```
import pandas as pd

airdata_url = 'https://raw.githubusercontent.com/jmstanto/\
    IDSpython/main/smallSurveyWithComments.csv'

airdata_df = pd.read_csv(airdata_url)

print(airdata_df.shape)
airdata_df.columns

(282, 32)
Index(['Destination.City', 'Origin.City', 'Airline.Status',
    'Age', 'Gender', 'Price.Sensitivity',...
'Likelihood.to.recommend','olong','olat','dlong','dlat',
'freeText'],
dtype='object')
```

The column list shows all of the variables we are used to working with as well as a final variable in the rightmost column called freeText. We will build a topic model based on the strings in freeText and use the document-topic vectors to correlate with another variable in the dataset: Likelihood.to.recommend. The shape of the data frame shows that there are n=282 rows and 32 variables.

Next, we will vectorize the text comments using exactly the same technique as shown earlier in the chapter. We begin by copying the string data out of the pandas DataFrame into a list so that our subsequent transformations leave the data frame intact.

```
sentlist = airdata_df["freeText"].to_list()

for i, sent in enumerate(sentlist):
    sentlist[i] = "".join([ch for ch in sent if ch not in
                    string.punctuation])

counter = CountVectorizer(min_df=10, max_df=0.9,
                    tokenizer=casual_tokenize,
                    stop_words='english')
```

```
countmatrix = counter.fit_transform(raw_documents=sentlist)

countmatrix.shape, type(countmatrix)

((282, 112), scipy.sparse.csr.csr_matrix)
```

We have increased the lower limit for a word to be included in out document term matrix: Across the 282 comments, a word must appear in at least 10 of them to be included in the resulting matrix. You can experiment with making this value larger or smaller. A larger value for min_df will make overall size of the vocabulary smaller, that is, fewer columns in our document-term matrix to represent the token counts.

In the next code cell, we will use a more systematic analysis of perplexity to choose an appropriate value of "k," the number of topics. As noted, there's no perfect number for this, and we always try to balance simplicity versus fit. More topics means finer gradations among the words associated with a topic.

```
import pandas as pd
from sklearn.decomposition import LatentDirichletAllocation

countdf = pd.DataFrame.sparse.from_spmatrix(countmatrix,
                    columns=counter.get_feature_names_out())

for i in range(2,20):
    print("k=", i, "topics:")
    lda_test = LatentDirichletAllocation(n_components=i,
            random_state=1, verbose=1, evaluate_every=5)
    lda_test.fit_transform(countdf)
```

The for loop at the end of this code block produces a lot of output. Using the arguments verbose=1 and evaluate_every=5, we get a view of the iterative fitting process that the LDA algorithm uses as well as a report of the perplexity value halfway through the fitting process and at the end. This output shows which models produce the lowest perplexity overall and which ones have the greatest reduction in perplexity between iteration 5 and iteration 10. As it turns out, the absolute lowest perplexity is for a simple k=2 model (perplexity=84.7). But we want to have something a little more interesting to look at as well as a few different vectors to test, so we will go for a k=4 solution with a final perplexity of 96.0. Even though this is slightly larger, it might be helpful to have more than two topics for this analysis. If we had used two topics, many words would be mushed together into each topic, and we probably would have had many interesting words like "flight" and "delays" occurring in the same topic with words like "little," "luggage," and "space."

Note that once we have settled on our preferred value of k, we need to rerun that model. This code also produces the graphic excerpted in Figure 13.4.

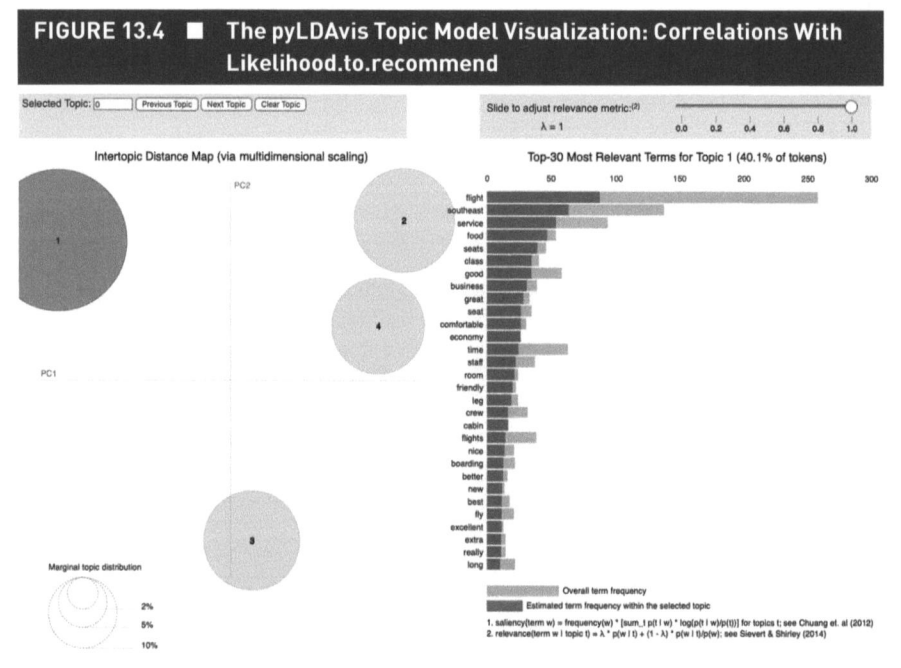

FIGURE 13.4 ■ The pyLDAvis Topic Model Visualization: Correlations With Likelihood.to.recommend

```
print("Repeating analysis at k=4:")

lda4air = LatentDirichletAllocation(n_components=4,
            random_state=1, verbose=1, evaluate_every=1)
lda4air.fit_transform(countdf)

pyLDAvis.lda_model.prepare(lda4air,countmatrix,counter)
```

Looking at the collection of words for the first topic, we see a list of words that includes "good," "great," and "comfortable." This looks promising, and we might hypothesize that document scores on this topic would have a positive correlation with the Likelihood. to.recommend variable.

We can and should view some of the more detailed numeric values to orient ourselves to what the analysis produced. We can use similar code as was presented earlier in the chapter:

```
for a, b in zip(list(counter.get_feature_names_out()),
                lda4air.components_[1]):
  if b > 14:
    print(a, b)
```

```
best 14.421630022917302
better 15.673967710984668
boarding 15.89888457239129
business 38.595163775844604
cabin 20.239177554467027
class 43.309037039602075
comfortable 32.709480140268454


doc_topic = lda4air.transform(countdf)

np.corrcoef(airdata_df["Likelihood.to.recommend"],
       [d[1] for d in doc_topic])

array([[1.      , 0.36817005],
    [0.36817005, 1.      ]])
```

This code cell e shows a truncated portion of the output from the for loop. These results, for words like "best," "better," and "comfortable," show the expected frequencies for each word. As shown earlier in the chapter, a higher value indicates that the word is more prevalent in that particular topic. We filtered out all words with expected frequencies of 14 or lower not because they are unimportant but rather to save space in the display. The output from np.corrcoef() at the end of the code block shows a value of r=0.37 between topic 1 and Likelihood.to.recommend. Looking over the words in this topic, it is evident that business class fliers like the comfort of their cabin as well as the opportunity to board earlier.

You could push this part of the analysis a little further by creating a linear regression model where you predict the Likelihood.to.recommend variable from more than one of the document-topic vectors. If you use this example, where we modeled four topics, make sure to use only three of the four document-topic vectors in your analysis. You may remember that for a given document, all of the document-topic scores add up to constant. This also means that any three of the columns can be used to exactly calculate the remaining column. If we tried to use all four in a prediction equation, we would create a problem in the linear regression model.

The comments on this dataset, and others like them, have the potential to indicate customer experiences that could have a plausible effect on LTR and NPS. This example shows how it would be possible to use topic modeling as a way of directing our attention to customer comments that are indicative of problems with satisfaction. As suggested earlier in the chapter, we could also explore sentiment analysis with our survey responses, although in one sense we already have a kind of sentiment score with our Likelihood.to.recommend variable. What we could also do would be to create a custom sentiment dictionary from our customer comments and the other data we have in this dataset. Then in the future we could use the keywords from that dictionary to quickly identify problems with certain routes or passenger experiences.

CHAPTER CHALLENGES

1. Read in a different text file from the web as shown in this chapter. For example, you could read in the text of the of a speech by another suffragist, Elizabeth Cady Stanton, from this page: http://www.historyplace.com/speeches/stanton.htm. Note that if you encounter an error, check that the URL has http and not https. Depending upon the setup of your notebook environment, your Python notebook may not be able to read an https file.

2. Process the new example from the previous exercise using BeautifulSoup to filter out the HTML tags. Check the results, and comment on any anomalies you see.

3. Use the appropriate Python code to organize your web page text by sentences. Report how many documents/sentences are in your corpus.

4. Use VADER or another sentiment analysis library of your choice to generate sentiment scores for each sentence in your corpus. Show the sentence with the highest sentiment score and the sentence with the lowest sentiment score.

5. Convert your corpus into a document-term matrix using the appropriate vectorizer. Report how many terms and documents the resulting data structure contains. List the 10 most frequently occurring words.

6. Create a word cloud from your document-term matrix.

7. Transform your data as needed to compute an LDA model using the sklearn library. Choose a four-topic model to start and feel free to change that number to improve your result. Generate an interactive graphic using the pyLDAvis library. Take a screenshot of the word distribution for the topic that looks most interesting to you.

8. For the topic you described in the previous problem, find and show the text of the document that best represents your favorite topic. Reinterpret your topic now that you have looked at a representative sentence.

9. Correlate a document-topic vector from your topic model with the sentiment scores generated from exercise 4.

14 IN THE SHALLOWS OF DEEP LEARNING

LEARNING OBJECTIVES

Describe the impact deep learning has had on the field of machine learning.

Explain how neural networks conceptually work.

Experiment with simple neural networks.

Explain key deep learning concepts.

THE IMPACT OF DEEP LEARNING

"Deep learning" is a term that refers to the use of layered neural networks to solve complex computational problems. Deep learning has had, and will continue to have, a huge impact on our lives. For example, it is the key machine learning technique that has enabled the development of machine translation, human-like conversational engines, generative artificial intelligence, and self-driving cars. Although the basic ideas that underpin deep learning date back to the 1960s, the field started to take off around 1999, when researchers began using graphics processing units (GPUs) to run some of the complex and time-consuming calculations that deep learning algorithms require.

To illustrate the power of deep learning, let's examine its impact on speech recognition. Before deep learning came along, computer applications could achieve only about 80 percent accuracy for speech recognition. This level of accuracy was far below what a human listener could do, which is around 95 percent accuracy. The computer level of accuracy did not materially change until around 2013, when deep learning was first used to address speech recognition. Accuracy then rapidly improved to the point that by 2017, deep learning algorithms could achieve essentially the same level of accuracy as humans. It is this machine learning technology that has enabled speech recognition to be used in apps such as Siri and Alexa as well as in the voice dictation available in your phone and/or laptop.

Image recognition is another domain that has demonstrated the power of deep learning. Prior to 2012, having a computer system recognize a dog or a cat in a digital photo was both difficult and error prone. However, similarly to the speech recognition example, there was a rapid decrease in error rates for identifying what was in an image starting around 2012, when deep learning started to be used. As our ability to use deep learning improved, deep learning systems have now become better than humans at identifying an object in an image.

HOW DOES DEEP LEARNING WORK?

In a previous chapter we discussed linear regression, a predictive technique to use input variables to predict a metric output variable. For example, we can use the weight of a vehicle and its engine horsepower to predict its fuel economy. After that, when we examined supervised machine learning techniques, we also examined two statistical classifiers, which can use input variables to predict a categorical outcome, such as whether a bank customer will default on a loan. Deep learning can address both regression and classification problems. To illustrate with a standard classification problem, think through an example of recognizing numeric digits from digital images. This is an important application in postal mail and shipping, where using an algorithm to recognize handwritten digits on a parcel make delivery more efficient. We can use supervised machine learning (with deep learning or another algorithm) to examine thousands of examples of images that contain a digit. In this scenario, each digit can be represented by a rectangular grid of black and white pixels. Thousands of these rectangular girds are processed through the deep learning model as training data. The fitted model can then be used to indicate which numeric digit is most likely when a new grid of pixels is presented.

Deep learning works by creating a network of artificial neurons. Each artificial neuron receives one or more inputs and produces an output in response. The most interesting aspect of an artificial neural network is that each neuron can learn from training examples. As each neuron learns, the pattern of outputs it produces in response to a set of inputs will change to become more accurate. A system with only one neuron is pretty boring—in fact, it is a little bit like a simple linear regression prediction model. The power of deep learning comes from connecting neurons together. A group of neurons that work in parallel is called a "layer," and a deep learning algorithm generally has input layers, hidden layers, and output layers. When a deep learning model has more than one hidden layer, it begins to have the potential to tackle some challenging regression and classification problems.

Let's explore the basic building block of a deep learning network—the neuron. Each neuron describes a calculation process with one or more input connections, a weight/importance value for each input, a transfer/activation function that combines the inputs in some way, and an output. Let's create a simple one-neuron model to help predict which car to buy. The neuron has three inputs to determine the output of the neuron. The neuron will produce an output value combining the three inputs. Table 14.1 lists the three inputs to our neuron and the weight/importance of each one (how the different factors weighted inputs combine) to produce a result on a scale of zero to 10.

So, for example, a car with 32 mpg (which means that X1=1), 190 hp (which means that X2=0), and an automatic transmission (which means that X3=1) would create an output of 8 (1*3 + 0*2 + 1*5 = 8). Figure 14.1 shows an example of the activation function that we could use for our neuron: The activation function outputs the value of 1 if the sum of the weighted inputs is greater than or equal to five or the value 0 if the sum is less than 5. If the output of the neuron equals 1, then it would be a good idea to buy the car!

A system with one neuron is limited in what it can accomplish, but as we add more neurons to a layer, we can get an improvement in predictive capability. However, neural networks with one layer can handle only the simplest of regression and classification problems—basically those situations in which the outputs are more or less a direct function of the inputs. In contrast, a multilayer network can solve more difficult prediction problems. In this situation, the networks can learn complex dependencies among the inputs that may lead to different

TABLE 14.1 ■ Three Inputs to a Neural Network Node		
Input	Input description	Weight
X1	Is the miles per gallon (mpg) greater than 30 mpg?	W1 = 3 (high MPG is important)
X2	Is the horsepower above 200?	W2 = 2 (high horsepower is slightly important)
X3	Is the transmission type auto?	W3 = 5 (an automatic transmission is very important)

FIGURE 14.1 ■ **Diagram of a Neuron, Showing the Rule Used by the Activation Function**

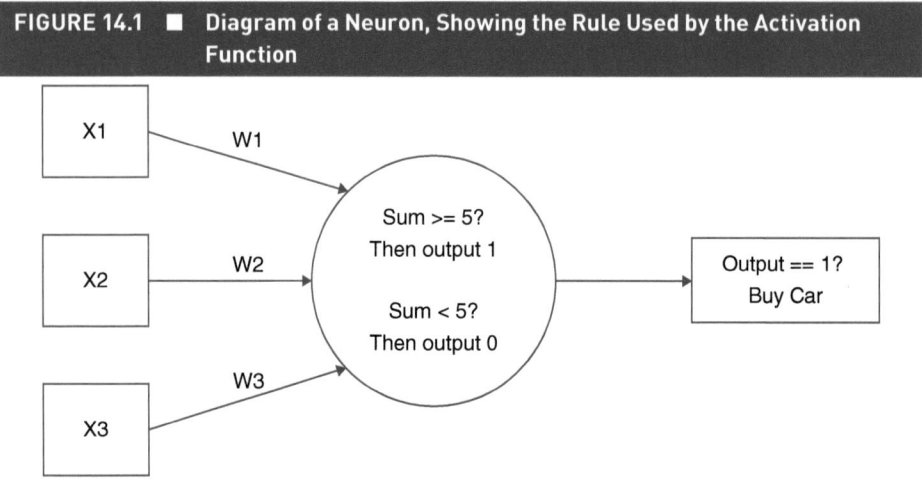

Note. From *Data Science for Business With R*, by J. S. Saltz and J. M. Stanton, p. 357. Copyright 2022 by Sage Publications. Reproduced with permission.

decisions about the output. To give a simple example, let's say we were trying to use deep learning to identify pictures of dogs. Perhaps we could begin to address that problem with a one-layer model. The algorithm might look for pointy ears, fur, four paws, and a tail. This simple, one-layer model might make a mistake, however, when presented with a picture of a large cat, which also has pointy ears, fur, four paws, and a tail. To accurately differentiate between a small dog and a large cat, we might need to add more layers of neurons to our deep learning model. These additional layers could process combinations of features that help distinguish a cat from a dog, such as the typical shape and length of a dog's nose.

We have examined one neuron and imagined several neurons working together in a layer, but how does the training process actually change the weights for each neuron? The answer is a procedure known as "backpropagation." Backpropagation adjusts the weights of a neural network based on the errors that occur in a batch of training examples. The training process divides the training data into smaller parcels called "batches," where the errors are accumulated during a batch and then used to tune the model weights when the batch is complete.

The reason the tuning activity is called backpropagation is that changing the weights actually begins by comparing the results from the output layer to the true values that have been supplied with the training data. After adjusting the weights for the final layer, we move backward to the layer that precedes the final layer, repeating the adjustment process there. After that we keep going backward through all of the other layers in the model, making adjustments as needed. We keep running training batches until we reach the point where the model does not seem to be improving. As with all supervised machine learning models, we must be careful not to overfit, or we will create a predictive model that works fine with the training data but that will not generalize to new data. You may remember from earlier chapters that we can hold out some testing data or use a k-fold cross validation strategy to prevent overfitting.

DEEP LEARNING IN PYTHON—A BASIC EXAMPLE

Training a deep learning network to identify an object in an image can take a long time because the training model needs to process millions of images that are paired with correct identifications. Part of the reason that deep learning networks have become more practical is the decreasing cost of specialized processing chips known as GPUs These GPUs were originally developed to help process images for gaming applications and other software that has intensive display needs. One of the tasks a GPU is particularly good at is rapidly computing arithmetic functions, like multiplication, over long lists of numbers. That's also one of the tasks needed to train a deep learning model. Although a simple neural network model can be trained on a regular laptop or desktop computer, the large models that create chatbots or perform machine translation need specialized systems with one or more GPUs, or else the training will take so long as to be impractical.

To demonstrate deep learning, we will use two Python packages developed at Google, known as Keras and Tensorflow. The word "keras" is Greek for "horn" and is also a literary reference about announcing the future—any apt idea for a powerful predictive method. Keras serves as the programming interface to Tensorflow, a package of deep learning modules, each of which implements a specialized type of neural network layer. Let's start with a simple example that addresses a regression problem:

```
# X is the weight of the vehicle in tons
X = [2.620, 2.875, 2.320, 3.215, 3.440, 3.460, 3.570, 3.190, 3.150,
3.440, 3.440, 4.070, 3.730, 3.780, 5.250, 5.424, 5.345, 2.200, 1.615,
1.835, 2.465, 3.520, 3.435, 3.840, 3.845, 1.935, 2.140, 1.513, 3.170,
2.770, 3.570, 2.780]

# Y is the fuel economy in mpg
Y = [21.0, 21.0, 22.8, 21.4, 18.7, 18.1, 14.3, 24.4, 22.8, 19.2, 17.8,
16.4, 17.3, 15.2, 10.4, 10.4, 14.7, 32.4, 30.4, 33.9, 21.5, 15.5, 15.2,
13.3, 19.2, 27.3, 26.0, 30.4, 15.8, 19.7, 15.0, 21.4]

import matplotlib.pyplot as plt

plt.scatter(X, Y)
```

Another automotive example! Here we have a small excerpt from the widely used "mtcars" dataset containing n = 32 observations of a predictor—the weight of a vehicle in tons—and the outcome variable, miles per gallon (mpg). Figure 14.2 shows a scatterplot of these two variables.

Figure 14.2 suggests a clear linear relationship between vehicle weight and fuel economy, with a negative slope, and a Y-intercept that probably lies somewhere close to 35 mpg. This is an analysis that we can easily tackle with a simple linear regression model (as discussed in Chapter 10). In fact, before we set up a neural network model to accomplish this, let's find out what the linear regression results would be:

FIGURE 14.2 ■ Scatterplot of Vehicle Weight (Tons; X) Versus Fuel Economy (MPG; Y)

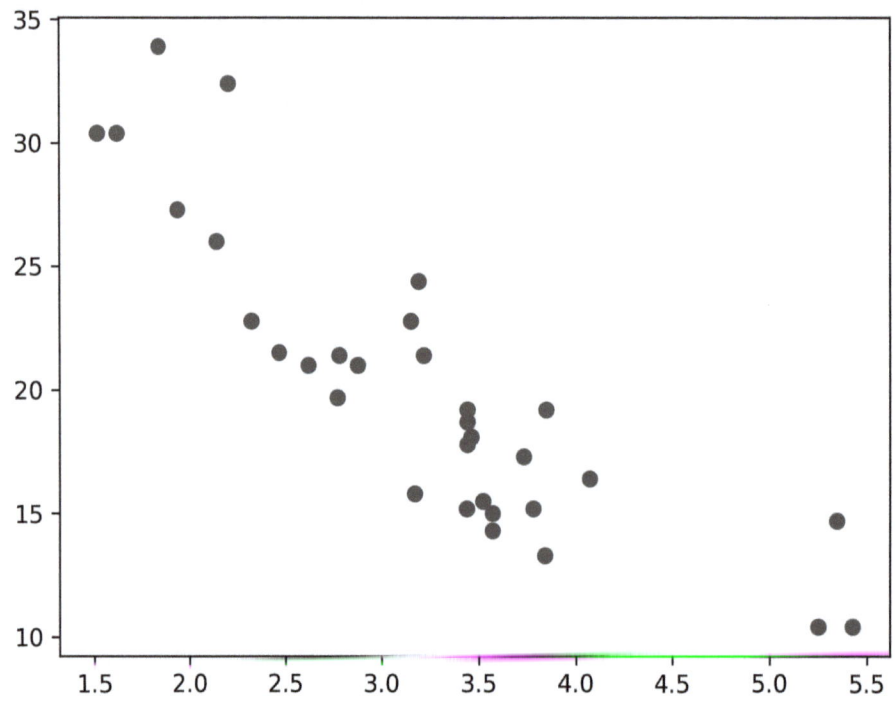

```
from sklearn.linear_model import LinearRegression
import numpy as np
regression = LinearRegression() # Instantiate the class
regression.fit(np.array(X).reshape(-1,1), Y)
regression.coef_, regression.intercept_
(array([-5.34447157]), 37.28512616734204)
```

Results of the regression analysis show a slope of –5.34 and a Y=intercept of 37.29. A supplemental line of code in the notebook for this chapter shows an R-squared value of 0.75. Now we will use Keras and Tensorflow to create the same model using a single neuron:

```
import tensorflow as tf
import keras
from keras import Sequential
```

```
model = tf.keras.Sequential()

model.add(tf.keras.layers.Dense(1, input_shape=[1]))

type(model)

keras.engine.sequential.Sequential
```

After importing the modules we need, we instantiate a sequential Keras model. Sequential in this case simply indicates that we can create as many layers as we want in the order that they will process the data. We used model.add() to accomplish this—in this case adding a single dense layer with one neuron. By default, that neuron will use a linear activation function, which essentially means we have created neural node that does the same thing as simple regression.

"Dense" is a deep learning term meaning "fully interconnected" and is an easy way of creating a set of neurons where each neuron is connected to all of the elements of the previous layer. In this case, because we are using one neuron, there isn't any connecting to do. The last line of code asks for a report of what type model is, and we learn that it is an instance of the keras.engine.sequential.Sequential class.

With simple regression, you might remember from Chapter 10 that there is a standard strategy for computing the result, and that strategy uses what is known as "ordinary least squares criterion"—we compute a slope and intercept that minimize the mean squared error of the data points from the fitted regression line. Neural networking offers many different computational strategies, so we have to make our choice explicit, as the next code cell shows:

```
model.compile(loss= "mean_squared_error", \
              optimizer=tf.keras.optimizers.SGD(0.04))

model.summary()

Model: "sequential"
_____
Layer (type)                    Output Shape              Param #
=================================================================
dense (Dense)                   (None, 1)                 2
=================================================================
Total params: 2
Trainable params: 2
Non-trainable params: 0
```

The first line of code "compiles" the model—that is, it completes the setup—by specifying mean squared error as the loss function to be used in the training process. This step also adds an optimizer that tells Tensorflow how to do the backpropagation. The keyword here is "SGD," which stands for stochastic gradient descent: a method for finding better choices for our slope and intercept at each stage of the training process. The value 0.04 is known as the "learning rate," and it controls how rapidly the model learns from its mistakes. The second line of code asks for a report of the structure of the model, and the output shows that we have a single node that outputs one value, and that has two trainable parameters—the slope and the intercept. Now we are ready to train the model:

```
fit = model.fit(X, Y, batch_size=8, epochs=200)

Epoch 1/200
4/4 [==========================] - 0s 4ms/step - loss: 153.9880
. . .
Epoch 200/200
4/4 [==========================] - 0s 5ms/step - loss: 8.7838
```

The fit() method conducts the training, using X and Y as the training data, treating groups of eight points as a batch for training purposes and running the training loop 200 times (epochs). The output here shows the value of the loss function for the first and last steps, and it is easy to see that the training process has reduced the mean squared error (loss) by a substantial amount. Let's find out how close we have gotten to the results produced by simple regression:

```
model.weights

[<tf.Variable 'dense/kernel:0' shape=(1, 1) dtype=float32,
numpy=array([[-5.173133]], dtype=float32)>,
    <tf.Variable 'dense/bias:0' shape=(1,) dtype=float32,
numpy=array([36.976517], dtype=float32)>]
```

There's a whole lot of technical information in there about how the two coefficients are stored, so we have highlighted the slope and intercept values in bold. Note that when you run the code in the notebook, you might get slightly different results because of the random initialization that Tensorflow uses at the start of the training process. If you compare these values to the results from the simple regression, you will see that both the slope and intercept are similar to those for the best-fitting line. If we continued the training process, we could get even closer. A supplemental line of code in the notebook for this chapter shows that we have exactly matched the R-squared value of 0.75 that we got from the simple regression.

What we have accomplished so far is something like swatting a fly with a boulder: We've used a highly complex modeling technology to do something that we did previously with a much simpler and much less computationally intensive technique. And we achieved essentially identical results to the simpler technique. But the value of neural networks is not in their simplicity but in their potential to accomplish more difficult predictive tasks. We can show this idea conceptually, using the same data, by creating a more complex neural model:

```
complex_model = tf.keras.Sequential()
complex_model.add(tf.keras.layers.Dense(10,\
         activation='sigmoid', input_shape=[1]))

complex_model.add(tf.keras.layers.Dense(1, input_shape=[1]))

complex_model.compile(loss= "mean_squared_error", \
         optimizer=tf.keras.optimizers.SGD(0.02))

complex_model.summary()

Model: "sequential_1"
_____
Layer (type)                 Output Shape          Param #
=================================================================
dense_1 (Dense)              (None, 10)            20

dense_2 (Dense)              (None, 1)             11

=================================================================
Total params: 31
Trainable params: 31
Non-trainable params: 0
```

The complex model we created in the code cell here adds one layer of neurons before the single neuron that we used in the previous model. This new layer includes 10 neurons that use a nonlinear activation function called "sigmoid." A sigmoid curve is an S-shaped curve that can help model more complex patterns of data, particularly where the Y variable changes suddenly in response to a relatively small change in X. Note that whereas our simple one-node model had two trainable parameters (slope and intercept), this complex model has 31 trainable parameters. As such, it might be better at prediction than the simple model. Here's the result:

```
complex_fit = complex_model.fit(X, Y, batch_size=8, epochs=200)

Epoch 1/200
4/4 [==============================] - 0s 3ms/step - loss: 229.5897
. . .
Epoch 200/200
4/4 [==============================] - 0s 4ms/step - loss: 12.2230
```

Although the value of the loss function after the last training epoch is slightly larger than for the simple model, a computation of the R-squared value (shown in the notebook for this chapter) shows a substantial improvement to 0.82. So our more complex model makes better predictions—using the training data—than the simple model.

As we conclude this basic example, there are a few important ideas to keep in mind. First, using Keras and Tensorflow, we can quickly and conveniently bring to bear an immense amount of predictive power by configuring an appropriate set of neural network layers. At the same time, however, we must guard against overtraining (also called "overfitting") because that predictive power comes with the potential pitfall of only memorizing the training data. We encountered this problem in Chapter 11 with classifier models, and we addressed it by using a holdout sample for validation and/or by using k-fold repeated cross validation. In the more realistic deep learning example that follows, we use these methods to control the training process and cross-check to make sure we have not overtrained.

DEEP LEARNING USING THE MNIST DATA

Let's now try a real-world example. Specifically, given an image, we would like to predict what handwritten digit is in the image (think about a bank trying to determine what is written on a check). We will use the MNIST dataset, which is publicly available and easy to access. The MNIST data consists of 60,000 28 x 28 grayscale images of handwritten digits (check out the MNIST website (http://yann.lecun.com/exdb/mnist/ for more information). Figure 14.3 shows an example (see the notebook code):

According to the MNIST dataset, the digit that was intended here was five. We want to use the many thousands of examples in the MNIST data to train a Keras model to recognize any of the 10 digits from zero to nine. The code to do this training is similar to our previous example:

FIGURE 14.3 ■ The First Handwritten Digit in the MNIST Training Data

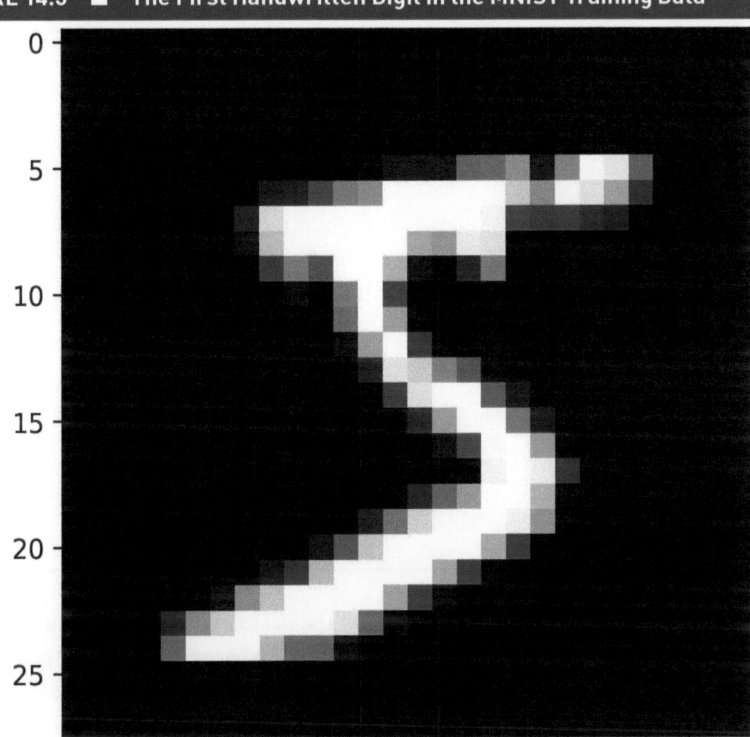

Note. The MNIST Dataset of Handwritten Digits (Images) by Yann LeCun (Courant Institute, NYU) and Corinna Cortes (Google Labs, New York), http://data.pymvpa.org/datasets/mnist/, licensed under CC BY-SA 3.0 DEED— attribution: ShareAlike 3.0 Unported https://creativecommons.org/licenses/by-sa/3.0/

```
(x_train, y_train),(x_test, y_test) = \
                tf.keras.datasets.mnist.load_data()
x_train_normalized = x_train / 255.0
x_test_normalized = x_test / 255.0

mnist_model = tf.keras.models.Sequential()
mnist_model.add(tf.keras.layers.Flatten(input_shape=(28, 28)))
mnist_model.add(tf.keras.layers.Dense(units=32,\
                                    activation='relu'))
mnist_model.add(tf.keras.layers.Dropout(rate=0.2))
mnist_model.add(tf.keras.layers.Dense(units=10, \
                                    activation='softmax'))

mnist_model.compile(optimizer=\
            tf.keras.optimizers.Adam(learning_rate=0.003),
            loss="sparse_categorical_crossentropy",
            metrics=['accuracy'])
```

```
mnist_model.summary()
Model: "sequential_4"

_____
Layer (type)                 Output Shape              Param #
=================================================================
flatten_2 (Flatten)          (None, 784)               0

dense_7 (Dense)              (None, 32)                25120

dropout_2 (Dropout)          (None, 32)                0

dense_8 (Dense)              (None, 10)                330
=================================================================
Total params: 25,450
Trainable params: 25,450
Non-trainable params: 0
```

The first line of code downloads the data, and the next two lines scale the greyscale pixel values to fractions between 0 and 1 (this helps the subsequent layers to process the graphical data more easily). After that we create a Keras sequential model, as we did earlier in the chapter.

Then we add four layers to the model. The first layer flattens out the 28 x 28 square grid of pixels into a long list of 768 numbers. Although there are more complex deep learning models that can handle a grid of data, we are using a simple model here. In the hidden layer, we use 32 neurons with the rectified linear unit (relu) activation function. The relu is a nonlinear activation function that produces an output value of zero if the value of the node was negative or the actual value of the node if that value is positive. A dropout layer randomizes which weights get tuned in each training epoch. Research has shown that this is an efficient method for preventing the training process from overfitting and developing a model that will generalize well to new data.

Finally, for our output layer, we use 10 neurons with the softmax function. This setup allows us to identify the probabilities of each of the 10 digits for each image we examine. For each training image, this layer will provide a list of 10 probability values that sum to one. The digit with the highest probability is the model's choice for that image. By having a complete list of 10 probability values, we can also tell how sure the model is about each choice.

The second-to-last line compiles the model. Note that we use "categorical_crossentropy" as our loss function: This is the appropriate choice for a multi-class prediction problem. A model summary shows the layers of the model and that we now have 25,450 weights and biases to train. That's more than 800 times as many as the model we trained earlier in the chapter! There are several implications of that large number: First, we have to have a lot of training data for backpropagation to be able to do its job on so many weights. The MNIST dataset has 60,000 instances, so we're covered there. Second, because there is so much "intelligence" in this model, we must guard against overtraining. That's why, when we loaded the data, we got both a training and a test dataset. We'll use the test dataset a little later to demonstrate that we did not overtrain, but we're also going to use a special strategy during training to keep an eye on the possibility of overtraining. Finally, keep in mind that even though we have more than 25,000 trainable weights in this model, many of the practical deep learning models currently in use have millions or billions of trainable weights. This is a small model in comparison to those.

Here's the code for training the model:

```
history = mnist_model.fit(x=x_train_normalized, y=y_train,
                          batch_size=4000, epochs=50,
                          shuffle=True, validation_split=0.2)

Epoch 1/50
12/12 [==========================] - 3s 133ms/step - loss: 1.6791 -
accuracy: 0.4781 - val_loss: 0.9624 - val_accuracy: 0.7906

. . .

Epoch 50/50
12/12 [==========================] - 0s 27ms/step - loss: 0.1789 -
accuracy: 0.9460 - val_loss: 0.1431 - val_accuracy: 0.9605
```

Once again, we have excerpted the output to show only the first and last epochs. We have reduced the loss and improved the accuracy in the training data. There's something new here, however. In the call to the fit() method, we specified "validation_split=0.2." That means that during each training epoch, 20 percent of the data are held back to cross-check how well the training process is going. At the end of each epoch, the training loss is reported as before, but we also get a report of the validation loss. If that validation loss starts to trend upward at any point in the training process, that means we have started to overfit. We either need to modify our model (e.g., to include more dropout layers), or we need to reduce the number of training epochs.

Earlier, when we compiled the model, we also asked Tensorflow to track model accuracy. As a result, at the end of every epoch, we also get a report of training accuracy and validation accuracy. Like keeping an eye on the loss function, you can also keep comparing the training accuracy and the validation accuracy at the end of every epoch. Ideally, validation accuracy will be quite similar to training accuracy. If, over a set of epochs, training accuracy keeps going up but validation accuracy starts going down, we know that we are overtraining. Figure 14.4 provides a graphical view of the progress of training and validation accuracy across the 50 epochs of our training process.

In Figure 14.4 we have a classic, positive example of a training process that is going well. Accuracy starts off low, about 50 percent, and goes up quite quickly, leveling off at about 100 epochs. During the last 40 epochs there are continual, small improvements in accuracy. Across all 50 epochs, validation accuracy consistently exceeds training accuracy by a small amount. We could probably get this model to become a little more accurate by running more epochs, but for demonstration purposes a final validation accuracy of 96 percent is quite good.

FIGURE 14.4 ■ Training and Validation Accuracy for the MNIST Model

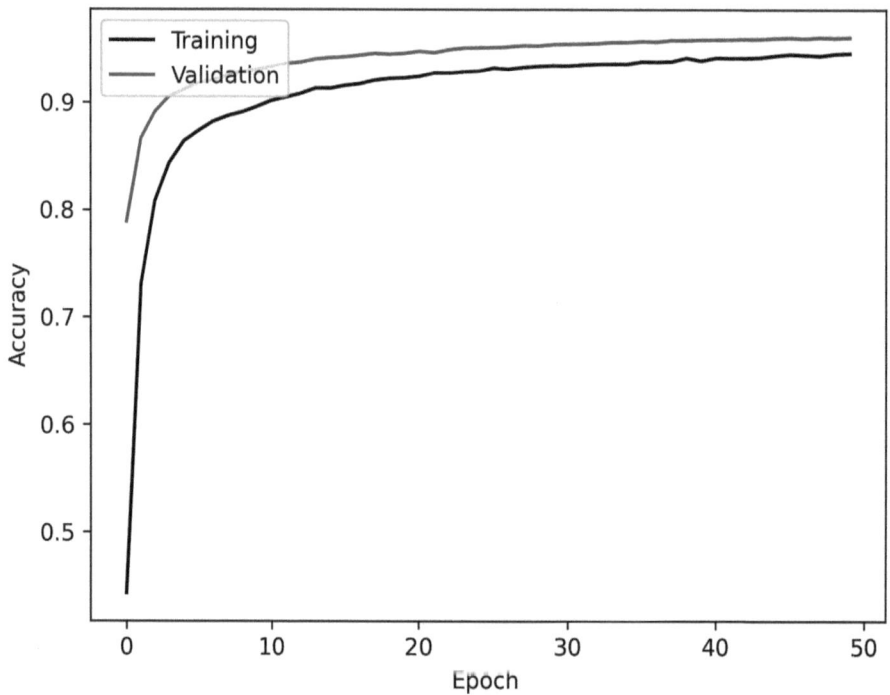

One last results check: Keras provides a convenient method called evaluate() that gives us a quick view of the accuracy of the model. We can run evaluate() on the held back test data as shown in this code cell:

```
mnist_model.evaluate(x=x_test_normalized, y=y_test, \
                    batch_size=4000)

3/3 [==============================] - 0s 17ms/step - loss: 0.1430 - accuracy: 0.9581

[0.1430191993713379, 0.9581000208854675]
```

The first line of output shows that the evaluate() method actually ran three batches of data through the trained model. That's because our test dataset had n = 12,000 instances. Deep learning specialists often call this process "inference," and it is conceptually identical to when we previously used the predict() method on a regression or classification model. The resulting loss and accuracy values are averaged across those three batches. The final line of output simply reaffirms what we already saw from the diagnostic output of the inference: a final model accuracy of 95.8 percent. Because this matches the final validation accuracy from the training process, we are satisfied that our model has been trained properly, not overtrained, and is capable of making good predictions of handwritten digits in the future.

This concludes our deep learning example for this chapter. The three models we demonstrated are all known as "feedforward" models because the data travels through the model in one direction only. The models we trained and tested also use a simple form of input data comprising a list of one or more X values that are presented to the model as a one-dimensional vector. More complex deep learning architectures exist that can handle multidimensional arrays of X values, such as full-color digital photos. A type of model called a "convolutional neural network" can examine small areas of these arrays looking for patterns that appear among neighboring data points. Another type of model, called a "recurrent" model, can be set up to process sequences of data such as text or audio. Finally, "transformer" models use deeply nested layers and rich cross-connections to detect complex interdependencies and linkages in data. These transformer models are the basis of applications such as machine translation and generative artificial intelligence.

CASE STUDY: USING DEEP LEARNING IN THE AIRLINE SURVEY

Let's apply our deep learning approach to our case study data. We will use our data to predict who are the detractors. We use the same steps as previously to read in the data and then transform the categorical variables into dummy codes.

```
survey['Detractor'] = \
        survey['Detractor'].replace({True: 1, False: 0})


features = ['Shopping.Amount.at.Airport', 'Age',
            'Price.Sensitivity', 'Airline.Status',
            'Type.of.Travel', 'Gender', 'Class']


dummy_df = pd.get_dummies(survey[features], sparse=False, \
                    drop_first=True)
dummy_df.columns

Index(['Shopping.Amount.at.Airport', 'Age', 'Price.Sensitivity',
    'Airline.Status_Gold', 'Airline.Status_Platinum',
    'Airline.Status_Silver', 'Type.of.Travel_Mileage tickets',
```

```
                 'Type.of.Travel_Personal Travel', 'Gender_Male', 'Class_Eco',
                 'Class_Eco Plus'], dtype='object')
```

We are ready to use deep learning, and the code will look similar to the examples we used earlier in the chapter. We can put aside 20 percent of the data for cross validation and use the rest for training:

```
from sklearn.model_selection import train_test_split

X = dummy_df
y = survey.Detractor

X_train, X_test, y_train, y_test = \
        train_test_split(X, y, test_size=0.25, random_state=11)

X_train.shape, X_test.shape, y_train.shape, y_test.shape

((66072, 11), (22024, 11), (66072,), (22024,))
```

The output here shows that we have 11 predictor columns. There are n = 66072 instances in our training data and n = 22024 instances in our test data. We can now set up a simple neural network model:

```
inp_size = X_train.shape [1]

survey_model = tf.keras.models.Sequential()


survey_model.add(tf.keras.layers.Dense(input_shape=[inp_size],\
                    units=33, activation='relu'))

survey_model.add(tf.keras.layers.Dropout (rate=0.2))

survey_model.add(tf.keras.layers.Dense (units=1, \
                    activation='sigmoid'))

survey_model.compile(optimizer=\
                    tf.keras.optimizers.Adam (learning_rate=0.006),
                    loss="binary_crossentropy",
                    metrics=['accuracy'])
```

```
survey_model.summary()

Model: "sequential_3"
_____
Layer (type)                 Output Shape              Param #
=================================================================
dense_85 (Dense)             (None, 33)                396

dropout_32 (Dropout)         (None, 33)                0

dense_86 (Dense)             (None, 1)                 34

=================================================================
Total params: 430
Trainable params: 430
Non-trainable params: 0
```

The first line of code figures out how wide the X data are by looking at the shape of X_train. After that we set up a neural model with one relu layer that sends its output to a single sigmoid neuron. The final layer is sigmoid because we have a binary outcome variable. For the same reason, the loss function is binary cross-entropy. We use one dropout layer between the relu and sigmoid layers to control overfitting. Finally, we can create and train our deep learning neural network:

```
history = survey_model.fit(x=X_train, y=y_train,
                           batch_size=6000, epochs=50,
                           shuffle=True, validation_split=0.2)

Epoch 1/50
9/9 [==============================] - 0s 19ms/step - loss: 0.4642
- accuracy: 0.8073 - val_loss: 0.4565 - val_accuracy: 0.8136
. . .
Epoch 50/50
9/9 [==============================] - 0s 15ms/step - loss: 0.4598
- accuracy: 0.8110 - val_loss: 0.4548 - val_accuracy: 0.8136
```

The final epoch shows a validation accuracy of 0.81, which is approximately the same as what we achieved with the classifier models we used in Chapter 11. With a simple neural model like this, we would not expect to achieve much if any improvement over a Naïve Bayes or decision tree classifier. We can run model_evaluate() to run the test set through the train model:

```
survey_model.evaluate(x=X_test, y=y_test, batch_size=6000)

4/4 [==============================] - 0s 4ms/step - loss: 0.4546 - accuracy: 0.8136
[0.4545832872390747, 0.813567042350769]
```

It is reassuring to find that the accuracy in the test set is identical to what we saw from the training process. Typically, we would also want to plot the training history to evaluate whether we should be running more or fewer epochs. It would also be possible to experiment with a more deeply layered model. Overall, we have learned that we can get a modest amount of predictive capacity out of a relatively small set of predictor variables using a simply neural net model.

CHAPTER CHALLENGES

1. We will use the classic iris dataset for these exercises. After importing sklearn.datasets, run my_data = sklearn.datasets.load_iris() to obtain the dataset. Run dir(my_data) to see all of the attributes of the data structure you imported from sklearn. Write a comment with your guesses about what each attribute contains.

2. Copy my_data.data into a new variable called X. Copy my_data.target into a new variable called y. Display the first and last instances of X and y, and report what you see.

3. Use plt.hist() from matplotlib to plot histograms of each of the four columns in X.

4. Use plt.hist() from matplotlib to plot a histogram of y.

5. Create a simple linear regression model that predicts y from the four columns of X. Note that linear regression is not the appropriate choice for this analysis because y is categorical—we are using it solely for convenience. Report on what you find.

6. Create a simple neural network model to predict y from X. Examine the code from this chapter's notebook. This is a multi-class problem: y has three class options labeled 0, 1, and 2. Create a model that has a single dense layer with three nodes and an input shape of [4] to accommodate the four X columns. Use softmax as the activation function. Write a comment saying why softmax is the best choice for this model.

7. Compile the model from step 6. Use the Adam optimizer with a learning rate of 0.002. The loss function should be "sparse_categorical_crossentropy," and the metric

should be ['accuracy']. Run model.summary() on the result, and report how many trainable weights the model has.

8. Train the model using X and y. Don't worry about validation data for this exercise. Use a batch size of 15 (10% of the data), 50 epochs, and shuffle=True. Report the final value of accuracy shown in the output.

9. Plot the accuracy history of the model training from exercise 8.

BIBLIOGRAPHY

Bassett, L. (2015). *Introduction to JavaScript object notation: A to-the-point guide to JSON*. O'Reilly Media, Inc.

Bennett, J. (2010). *OpenStreetMap*. Packt Publishing Ltd.

Blake, C., Stanton, J. M., & Saxenian, A. (2013). Filling the workforce gap in data science and data analytics. In *Proceeding of the Annual Meeting of the iSchools (iConference)*, Fort Worth TX.

Bulmer, M. (2003). *Francis Galton: Pioneer of heredity and biometry*. Johns Hopkins University Press.

Bureau of Labor Statistics, US Department of Labor. (2017). Statisticians. *Occupational Outlook Handbook* (2016–17 ed.). Retrieved May 30, 2017, from https://www.bls.gov/ooh/math/statisticians.htm

Chen, D. Y. (2018). *Pandas for everyone: Python data analysis*. Addison-Wesley.

Denis, D. J. (2021). *Applied univariate, bivariate, and multivariate statistics using Python: A beginner's guide to advanced data analysis*. John Wiley & Sons.

Dobson, A. J., & Barnett, A. G. (2018). *An introduction to generalized linear models*. CRC press.

Dodge, Y. (2008). *The concise encyclopedia of statistics*. Springer Science & Business Media.

DuBois, P. (2008). *MySQL*. Pearson Education.

Dunnington, W. (1927). The sesquicentennial of the birth of Gauss. *Scientific Monthly*, *24*(5), 402–414.

Fischer, H. (2011). A history of the central limit theorem: From classical to modern probability theory. In *Sources and studies in the history of mathematics and physical sciences*. Springer.

Friendly, M., & Denis, D. J. (2001). Milestones in the history of thematic cartography, statistical graphics, and data visualization. Retrieved 31 July 2023, from http://www.datavis.ca/milestones

Grimmett, G. R., & Stirzaker, D. R. (1992). *Probability and random processes* (2nd ed.). Clarendon Press.

Géron, A. (2022). *Hands-on machine learning with Scikit-Learn, Keras, and TensorFlow*. O'Reilly Media.

Hahsler, M., Hornik, K., & Reutterer, T. (2006). Implications of probabilistic data modeling for mining association rules. In M. Spiliopoulou, R. Kruse, C. Borgelt, A. Nuernberger, & W. Gaul (Eds.), *From data and information analysis to knowledge engineering, studies in classification, data analysis, and knowledge organization* (pp. 598–605). Springer-Verlag.

Hunt, J. (2019). *A beginner's guide to Python 3 programming*. Springer.

Idris, I. (2015). *NumPy: Beginner's guide*. Packt Publishing Ltd.

Khan Academy | Free online courses, lessons & practice. (n.d.). Khan Academy. http://www.khanacademy.org/

Kruskal, W. H. (1980). The significance of Fisher: A review of R. A. Fisher. The life of a scientist, by Joan Fisher Box. *Journal of the American Statistical Association*, *75* (372), 1019–1030.

Lee, H., & Song, J. (2019). Introduction to convolutional neural network using Keras; an understanding from a statistician. *Communications for Statistical Applications and Methods*, *26*(6), 591–610.

Mackenzie, D. (1981). *Statistics in Britain, 1865–1930: The social construction of scientific knowledge*. Edinburgh University Press.

McKinney, W. (2017). *Python for data analysis: Data wrangling with pandas, NumPy, and IPython* (2nd ed.). O'Reilly.

Müller, A. C., & Guido, S. (2016). *Introduction to machine learning with Python: A guide for data scientists*. O'Reilly Media.

Palanisamy, S., & SuvithaVani, P. (2020, January). A survey on RDBMS and NoSQL databases MySQL vs. MongoDB. In *2020 International Conference on Computer Communication and Informatics (ICCCI)* (pp. 1–7). IEEE Xplore

Pearson, E. S. (1990). *'Student', A statistical biography of William Sealy Gosset*. Oxford University Press.

Richert, W. (2013). *Building machine learning systems with Python*. Packt Publishing Ltd.

Rivas, P. (2020). *Deep learning for beginners: A beginner's guide to getting up and running with deep learning from scratch using Python*. Packt Publishing Ltd.

Saltz, J., & Heckman, R. (2016). Big data science education: A case study of a project-focused introductory course *Themes in science and technology education, 8*(2), 85–94.

Saltz, J. S., Heckman, R., Crowston, K., You, S., & Hegde, Y. (2019, January). Helping data science students develop task modularity. In *Proceedings of the Hawaii International Conference on Systems Science (HICSS)*, 1–10. Computer Society Press.

Saltz, J., Shamshurin, I., & Connors, C. (2017). Predicting data science sociotechnical execution challenges by categorizing data science projects. *Journal of the Association for Information Science and Technology, 68*(12), 2720–2728.

Saltz, J. S., & Stanton, J. M. (2017). *An introduction to data science*. Sage Publications.

Stanton, J. M. (2021). Evaluating equivalence and confirming the null in the organizational sciences. *Organizational Research Methods, 24*(3), 491–512.

Stanton, J. M. (2013). Data mining: A practical introduction for organizational researchers. In J. M. Cortina, & R. S. Landis (Eds.), *Modern research methods for the study of behavior in organizations*, 199–230. Routledge.

Stanton, J. (2015). Sensing big data: Multimodal information interfaces for exploration of large data sets. In S. Tonidandel, E.B. King, & J. M. Cortina (Eds.), *Big data at work: The data science revolution and organizational psychology* (pp. 158–177). Routledge.

Stanton, J. M. (2017). *Reasoning with data: An introduction to traditional and Bayesian statistics using R*. Guilford Publications.

Stanton, J. M., Kim, Y., Oakleaf, M., Lankes, R. D., Gandel, P., Cogburn, D., & Liddy, E. D. (2011). Education for eScience professionals: Job analysis, curriculum guidance, and program considerations. *Journal of Education for Library and Information Science, 52*(2), 79–94.

Stanton, J., Palmer, C. L., Blake, C., & Allard, S. (2012). Interdisciplinary data science education. In *Special issues in data management, American Chemical Society* (pp. 97–113). American Chemical Society.

Trappenberg, T. P. (2019). *Fundamentals of machine learning*. Oxford University Press.

Varoquaux, G., Buitinck, L., Louppe, G., Grisel, O., Pedregosa, F., & Mueller, A. (2015). Scikit-learn: Machine learning without learning the machinery. *GetMobile: Mobile Computing and Communications, 19*(1), 29–33.

Weisberg, H. F. (1992) *Central tendency and variability*. Sage University Paper Series on Quantitative Applications in the Social Sciences. Sage.

Ward, M., Grinstein, G. & Keim, D. (2015) *Interactive data visualization*. CRC Press,

Wikipedia contributors. (2022). E-Science. Retrieved August 15, 2022, from http://en.wikipedia.org/wiki/E-Science.

Wikipedia contributors. (2023a). E-Science librarianship. Retrieved July 31, 2023, from http://en.wikipedia.org/wiki/E-Science_librarianship.

Wikipedia contributors. (2023b). OpenStreetMap. Retrieved July 31, 2023, from https://en.wikipedia.org/wiki/OpenStreetMap.

INDEX

Abstraction, 60
Accuracy, 209
Activation function, 265, 266 (figure)
Adult dataset, 200, 201 (table), 236 (figure)
Affinity analysis, 225
Airline status, 123
Airline survey, 237–241, 239 (figure), 240 (table), 241 (figure)
Alexa, 264
AlphaDF, 68, 69
Analysis phase, 3
Antecedent, 226
Anthony, S. B., 244, 249 (figure), 251
API. *See* Application programming interface (API)
append() method, 31
Application programming interface (API), 117, 118
Apriori, 233
Arguments, 60, 61
Artificial neurons, 265
Association rules data, 225–226, 225 (figure)
 airline survey, 237–241, 239 (figure), 240 (table), 241 (figure)
 visualizing and screening, 235–237, 236 (figure)
 working, 233–235, 234 (table)
Association rules mining, 226–233
 baskets, 227
 columns_ method, 228
 frequently each education category occurs, 231, 232 (figure)
 Groceries dataset, 226

grocery items with a minimum value of 0.10, 229, 230 (figure)
Asymmetric distributions, 84
Attention to quality, 4

Backpropagation, 266
Barcode scanner, 2, 3
Base maps, 163
Bayes, T., 206
BeautifulSoup, 246
Becker, B., 200
Bell curve, 87
Bernoulli, J., 99
Best-fitting regression, 181, 182 (figure)
Big data, 7
Binary formats, 110
Bins, 83
Boolean variable, 51
Boxplots, 136, 137 (figure)

CalcNPS() function, 173
Cardano, G., 99
Car maintenance, 183–191
 cumulative cost, oil changes, 190
 finished regression model, 186
 NumPy array, 185
 plt.scatter() function, 184
 R-squared value, 187, 189
 scatterplot, 184, 185 (figure), 188 (figure)
Cartesian grid, 159
Case studies
 association rules, airline survey, 237–241, 239 (figure), 240 (table), 241 (figure)
 deep learning, 277–280

explore NPS by state and city, 172–177, 175 (figure), 176 (figure)
linear models, 194–197
LTR distributions, 88–90, 89 (figure)
Net Promoter Score (NPS), 20–21, 37–39, 70–73, 257–261, 260 (figure)
new treatment, impact of, 105–107
reading, cleaning, and exploring survey dataset, 53–56
supervised models, 216–220, 220 (figure)
survey dataset, 122–126
visualizing key attributes related to NPS, 146–154, 147 (figure)–149 (figure), 151 (figure), 152 (figure), 154 (figure)
Cells, 11, 34
Census, 115
Central limit theorem, 99–100
Chatbots, 244
choices(), 95, 96
Choropleth() function, 165
Choropleth map, 165, 166 (figure), 168 (figure), 170 (figure), 171 (figure), 175 (figure), 176 (figure)
Citi Bike program, 120
Classes, 67–70
Classifiers, 199–220
 detractors, 220 (figure)
 example, 200–205
 Naïve Bayes, 206–211
 supervised learning, 211–216
Coefficient of determination, 187

Cohen's Kappa accuracy, 209
Color-coding, 158
Comprehension, 48
Conditional evaluations, 34–36
Confidence, 226, 234 (table), 236 (figure)
Confusion matrix, 209
connect() method, 114
Consequent, 226
Conviction, 234
Convolutional neural network, 277
copy() method, 49
Correlation, 76
CountVectorizer, 253
Credit analysis, 6
Credit card processing, 110
CSV files, 42–44, 245
    census_df, 43, 44
    read_csv(), 43, 44

Darwin, C., 76
Data acquisition, 3
Data analysis, 3, 4, 52
Data architecture, 3
Databases
    accessing, 113–117
    engine, 114
    external, 112
    knowledge discovery, 246
    SQL, 117
    sqlite3, 113
Data cleaning, 53, 111
Data formats, 110
DataFrames, 17, 29
    accessing columns in, 30–33
    ages column, 31
    as argument, 63
    column-sorted, 63
    count, 28
    exploring, 28–30
    function, 27
    grouping within, 51–53
    max, 29
    mean, 28
    min, 29
    Net Promoter Score calculation, 37–39
    read_excel() function, 111
    reset_index(), 70
    sorting and grouping, 49–51
    specific rows and columns in, 33–34
    square brackets, 32
    statistical calculations, 78
    std, 29
    subsets with conditional evaluations, 34–36
    variables/columns of, 37
Data mining
    data preparation, 224
    exploratory data analysis, 224
    interpretation of results, 225
    model development, 224–225
Data munging, 41–56
    cleaning up the elements, 47–49
    CSV files, 42–44
    grouping within DataFrames, 51–53
    removing rows and columns, 44–45
    renaming rows and columns, 46–47
    sorting and grouping DataFrames, 49–51
Data problems, 5–7
Data science
    banks, 6
    definition of, 2–3
    retailers, 6
    skills, 4–5
    steps in, 3–4
Data screening, 53
Datasets, 110
    CSV, 111
    JSON, 117–122
    MNIST, 272–277
    survey, 53–56, 122–126
    training and cross validation, 191–192
Data storage, 110
Data transformations, 4, 131
DecisionTreeClassifier, 212, 213
Decision trees, 214 (figure)
Deep learning, 10, 263–280
    activation function, 265, 266 (figure)
    backpropagation, 266
    batches, 266
    case study, 277–280
    image recognition, 264
    impact of, 264
    MNIST data, 272–277, 273 (figure), 276 (figure)
    neural network node, 265, 265 (table)
    in Python, 267–272, 268 (figure)
    speech recognition, 264
    working, 264–266
Defensive coding, 64–67
Degrees of freedom, 79
Dense, 269
Describe function, 29
Descriptive statistics
    understanding, 77–80
    using, 80–83
Detractors, 220 (figure)
Dispersion, 29, 79
Distribution
    asymmetric, 84
    likelihood to recommend (LTR), 88, 89 (figure)
    normal, 87–88 88 (figure)
    sampling, 95, 100–102
    using histograms, 83–87, 85 (figure), 86 (figure)
Distribution of sampling means, 104, 105
Documentation, 11, 63
Doppler radar, 6
dtypes function, 44
Dummy coding, 206
Dynamically typed language, 64

Einstein, A., 76
Engine, 113
Ethical reasoning, 4
Eugenics, 76
Excel data, 111–112
Exchange rates, 118
execute() method, 114
External databases, 112

Facebook, 8
Family data, 24, 24 (table)
Feature engineering, 203
Feedforward models, 277
findall() method, 227

Fisher, R., 76
fit() function, 183, 270
Folium, 160–166, 162 (figure)
  base maps, 163
  choropleth map, 165, 166 (figure)
  default projection, 162
  implication, 162
  JSON files, 163
Folium.RegularPolygon-Marker(), 167
Frequency histogram, 83
  census population values, 83, 85 (figure)
  likelihood to recommend variable, 147 (figure)
  population with bin size of 5 million, 85, 86 (figure)
Frost, R., 252 (figure)
F1 score, 209
Functions, 59–73
  abstraction, 60
  anatomy of, 61, 61 (figure)
  classes and methods, 67–70
  create and use, 60–61
  defensive coding, 64–67
  definition of, 60
  Net Promoter Score, 70–73
  in Python, 61–64, 61 (figure)
  reusability, 60

Gall-Peters projection, 159, 160, 161 (figure)
Galton, F., 76
Gauss, C. F., 87, 181
Generative artificial intelligence, 277
Geocodes, 157, 158
Gini, 214
Google Colab, 11, 12, 16
Google Drive, 11
Gosset, W. S., 76, 77
Government datasets, 110
Graphics processing units (GPUs), 264, 267
Grinstein, G., 130
groupby() method, 51
Grouping DataFrames, 49–51

head() method, 49
help() function, 67

Histograms, 83–87, 85 (figure), 86 (figure), 98, 99
  implications of, 99
  likelihood to recommend across travel categories, 147, 148 (figure)
  using matplotlib, 99
Hosted notebooks, 16
HousingMaps, 158
HTTP GET request, 118
Human-readable formats, 110
Human visual system, 129–130
Hypertext markup language (HTML), 245, 246

Image recognition, 264
Impurity, 214
Independent samples t-test, 107
Inferential statistics, 77
Information visualization, 130
Interactive Data Visualization, 130
Intercept, 180, 186
Intuitive encoding, 131

Java Script Object Notation (JSON), 110, 117–122
  foreign exchange rates, 118
  queries, 118
  from URL, 121
Jupyter notebooks, 11–12, 63, 162

Kaggle notebooks, 16
Keim, D., 130
Keras, 267
k-fold cross validation, 193
Khan Academy, 7–8
Knowledge discovery, 246
Kohavi, R., 193

Latent Dirichlet Allocation (LDA), 253
Latitude lines, 158, 159 (figure)
Law of large numbers, 99–100
Leverage, 234
Likelihood to recommend (LTR), 88, 89 (figure), 123

frequency histogram of, 147 (figure)
histograms of, 148 (figure)
multiple boxplot of, 149 (figure)
Linear models, 179–197
  algebra, 182
  best-fitting regression, 181, 182 (figure)
  car maintenance, 183–191, 185 (figure), 188 (figure)
  case study, 194–197
  k-fold cross validation, 193
  Python, 183
  training and cross validation datasets, 191–192
  Y-intercept, 182
Linear regression, 200, 264
  matrix math, 182
  slope and intercept, 180
Lists
  comprehensions, 53
  creating and using, 12–15
  slicing, 15–16
loc() method, 31
Longitude lines, 158, 159 (figure)
Lowe, R., 132

Machine learning, 180, 203
Machine translation, 277
Magic, 82
Map projection, 159
Map visualizations, 157–177
  basics, 158–160
  with Folium, 160–166, 162 (figure)
  Gall-Peters projection, 160, 161 (figure)
  latitude lines, 158, 159 (figure)
  lines of longitude, 158, 159 (figure)
  Mercator projection, 159, 160 (figure)
  showing points on, 166–172, 168 (figure), 170 (figure), 171 (figure)
Massachusetts Institute of Technology, 42
Matplotlib, 17, 83, 95, 134

Matrix algebra, 180
Mean, 28, 29, 77
Median, 29
median_pop, 51, 77
Mercator projection, 159, 160 (figure)
Metadata, 24
Methods, 67–70
Microsoft Excel spreadsheets, 110
MNIST data, 272–277, 273 (figure), 276 (figure)
Mode, 29, 77
Models
  definition of, 180
  linear. *See* Linear models
Moore's Law, 110
Munging. *See* Data munging
my_classification_report(), 211
my_family DataFrame, 32. *See also* DataFrames
Naïve Bayes
  in Python, 206–211
  supervised learning with, 205–206
Natural language processing, 246
Natural Language Toolkit (NLTK), 248–250
Nature versus nurture, 76
Net Promoter Score (NPS), 20–21, 257–261, 260 (figure)
  calculation, using DataFrames, 37–39
  function, 70–73
  pivot table grouped by age and travel type, 154 (figure)
  by state and city, 172–177, 175 (figure), 176 (figure)
  for various age groups, 151 (figure)
  visualizing key attributes related to, 146–154, 147 (figure)–149 (figure), 151 (figure), 152 (figure), 154 (figure)
Neural network node, 265, 265 (table)
Nominatim geocoding service, 167

Normal distribution, 87–88
  bell shape of, 87
  definition of, 83
  random, 87, 88 (figure)
Normalization, 246
normalvariate() function, 87
Not a number (NaN), 54
numberize(), 80, 81
Numeric Python, 17
NumPy, 17

Object-oriented programming, 67
One-hot encoding, 206
OpenDocument format, 110
OpenStreetMap, 167
Ordinary least squares criterion, 182, 269
Overtraining, 272

Pandas DataFrame, 17, 25–27
  data analysis, 52
  dropna() method, 54
  dummy coding, 207
  function, 62
  JSON, 120
  Series, 101
  sorting and grouping capabilities, 53
  statistical mode function, 82
  subsets of rows and columns, 53
  to_sql(), 113
Parameter, 61
Pareto distribution, 87
Pareto, V., 85, 87
pd.DataFrame() function, 27
Pearson, K., 76
Perplexity, 253
plt.scatter() function, 184
Point-of-sale system, 3
Polygon data, 163
Population, 77, 94, 95
Precision, 209
Price sensitivity, 123
Problem identification, 6
Problem-solving, 6
Proprietary format, 110
Punkt, 249
pyLDAvis, 255, 255 (figure), 260 (figure)

Python, 3, 7, 9–21
  basic plots in, 132–134, 133 (figure), 134 (figure)
  deep learning in, 267–272, 268 (figure)
  extensibility of, 10
  Folium library, 160–166, 162 (figure)
  function, 59–73
  in Jupyter notebook, 11–12
  linear model, 183
  lists, 12–15
  Naïve Bayes in, 206–211
  Net Promoter Score (NPS), 20–21
  object-oriented capabilities, 68
  Package Index, 17–20
  ready to use, 10–11
  rows and columns, 23–39
  sample plot, 19 (figure)
  sampling in, 95–96
  SciKit-Learn package, 180
  slicing lists, 15–16
  SQL, 113
  standard deviation, 79
  variance, 79
  virtual machine, 16–17
Python Software Foundation, 11

Quartiles, 37, 102

Random normal distribution, 87, 88 (figure)
random.seed(), 96
Range, 78
read_census() function, 81
read_excel() function, 111
Recall, 209
Recurrent model, 277
Regression, 76, 180
Regression trees, 211–216
Removing rows and columns, 44–45
Renaming rows and columns, 46–47
RepeatedKFold() function, 193
Repetitive sampling process, 96–99, 98 (figure)

Index

replace() method, 80, 205
reset_index() method, 30
Return values, 61
Reusability, 60
Reverse-J distribution, 85
reverse() method, 69, 70
Risks, uncertainty and, 6
Rothhamsted Experimental Station, 76
round() method, 82
Rows and columns, 23–39. *See also* DataFrames
 exploring DataFrames, 28–30
 pandas DataFrame, 25–27
 removing, 44–45
 renaming, 46–47
 two-dimensional representation of, 24, 24 (table)
R-squared value, 187, 189, 272
rURL, 119

sample_means, 101
Sampling, 94 (figure)
 definition of, 94
 population, 77
 in Python, 95–96
 repetitive, 96–99, 98 (figure)
 with thresholds, 102–105
Sampling distribution, 95, 100–102
Satellite, 169
Scattergram, 184. *See also* Scatterplot visualizations
Scatterplot visualizations, 141–145
 key_state, 143
 with key states labelled, 144, 144 (figure)
 variables, 142, 142 (figure)
SciKit Learn (sklearn) package, 183
sDF, 30
Seaborn, 132, 134–141
 attributes, 135
 bar chart, 140
 bar plot, 140, 141 (figure)
 boxplots, 136, 137 (figure)
 data, 135
 geometry, 135

histplot() function, 135
line plot, 139, 140 (figure)
percentage change, positive *vs.* negative growth states, 138, 139 (figure)
population values, 136 (figure)
Seaborn, S. N., 132
Sentiment analysis, 248–251
Sigmoid, 271
Siri, 264
Skewness, 98
Sklearn, 17
Slicing, 15–16
Smart devices, 7
Sorting DataFrames, 49–51
Speech recognition, 264
SQL. *See* Structured query language (SQL)
Sqlalchemy, 113, 114, 116
Sqlite3, 113
Square brackets
 conditional expression, 35
 in pandas, 32
Squared deviations, 79
Stamen maps, 170
Standard deviation, 78, 79, 104
Statistical inference, 6, 103
Statistical learning, 180
Statistics, 77
 descriptive, 77–80
 distribution. *See* Distribution
 inferential, 77
Storage, 109–126
Stratified k-fold cross validation, 193
Structured query language (SQL), 110, 113
 accessing data, 117
 connection, 114
 keywords and syntax, 115
 queries, 114, 116, 117
Student's t-test, 76
Subject matter experts (SMEs), 5
sum() function, 38
Supervised learning, 180–181, 199–200
 classification and regression trees, 211–216
 classifiers, 200

definition of, 180
with Naïve Bayes, 205–206
Supervised machine learning, 264
Supervised techniques, 224
Support, 226, 234 (table), 236 (figure)
Survey dataset, 53–56, 122–126

Tags, 245
tail() function, 46
Tensorflow, 267
Terrain maps, 169
Test dataset, 192
text(), 113
Text analysis, 243–261
Text data, 244
Text files, reading in, 245–246
Text mining, 246, 257
Tokens, 246
Topic modeling, 251–257, 252 (figure), 255 (figure)
train_test_split() function, 192
TransactionEncoder(), 227, 234
Transformer model, 277
Translation mechanisms, 131
True and Falses, 34, 35
TypeError, 66

Uncertainty, risk and, 6
Uniform resource locator (URL), 43
University of California, Irvine (UCI) Machine Learning repository, 200
Unstructured data, 244
Unsupervised learning, 180–181
Unsupervised techniques, 224
U.S. Census Bureau, 42
U.S. Geological Service, 169

Valence Aware Dictionary and sEntiment Reasoner (VADER), 248–251
value_counts() function, 203
Variance, 78
 calculation of, 79, 79 (table)
 compute, 79

Vector graphics, 169
Virtual machine, 16–17
Visualization, 4, 244
    adult dataset, 236 (figure)
    aggregates, 132
    avoid overplotting, 131
    case study, 146–154, 147 (figure)–149 (figure), 151 (figure), 152 (figure), 154 (figure)
    color palette, 132
    comparison across panes, 132
    confidence, 236 (figure)
    connections, 132
    data transformations, 131
    definition of, 130
    information, 130
    intuitive encoding, 131
    map, 157–177
    overview, 130–132
    patterns *vs.* details, 131
    scatterplot, 141–145
    simplicity, 131
    support, 236 (figure)
    translation mechanisms, 131
    using seaborn, 134–141
    value and axis ranges, 131
Visual pattern recognition, 129

Ward, M., 130
W3C consortium, 110
Web address, 43
The West Wing, 132
Wikipedia, 7
Word cloud, 246–248, 249 (figure)

XML, 110

If you are based in the European Union and have safety concerns related to physical Sage products, in compliance with GPSR, please contact our Authorized Representative:

International Associates Auditing & Certification Limited
The Black Church, St Mary's Place,
Dublin 7, D07 P4AX Ireland
EUAR@ie.ia-net.com